Exercises in
PHYSICAL
GEOLOGY

Eleventh Edition

W. KENNETH HAMBLIN

Brigham Young University

JAMES D. HOWARD

Prentice
Hall

PRENTICE HALL, Upper Saddle River, NJ 07458

Library of Congress Cataloging-in-Publication Data

Hamblin, W. Kenneth (William Kenneth)
    Exercises in physical geology / W. Kenneth Hamblin, James D.
  Howard. — 11th ed.
      p.   cm.
    ISBN 0-13-062090-4
    1. Physical geology—Laboratory manuals. I. Howard, James D.
  II. Title.
  QE28.2.H36 2001
  551—dc21
                                    2001036254
                                          CIP

Acquisitions Editor: Patrick Lynch
Editor-in-Chief: Sheri L. Snavely
Production Editor: Debra A. Wechsler
Assistant Vice President of Production and
   Manufacturing: David W. Riccardi
Executive Managing Editor: Kathleen Schiaparelli
Senior Marketing Manager: Christine Henry
Manufacturing Buyer: MIchael Bell
Manufacturing Manager: Trudy Pisciotti
Director of Design: Carole Anson
Art Director and Cover Designer: John Christiana
Managing Editor, Audio/Visual Assets: Grace Hazeldine
Associate Editor: Amanda Griffith
Director of Creative Services: Paul Belfanti
Editorial Assistants: Nancy Bauer; Sean Hale
Copy Editor: Jocelyn Phillips
Composition: Vicki L. Croghan, Envision Inc.
Project Coordinator, Artworks: Connie Long
Senior Manager, Artworks: Patty Burns
Production Manager, Artworks: Ronda Whitson
Cover Photo: Volcanic features in the Western Grand Canyon between
Toroweap Valley and Whitmore Wash.. *(W. Kenneth Hamblin)*

PRINTED IN THE UNITED STATES OF AMERICA

10 9 8 7 6 5 4 3 2 1

ISBN 0-13-062090-4

Pearson Education LTD., *London*
Pearson Education Australia PTY, Limited, *Sydney*
Pearson Education Singapre, Pte. Ltd.
Pearson Education North Asia Ltd., *Hong Kong*
Pearson Education Canada, Ltd., *Toronto*
Pearson Educación de Mexico, S. A. de C. V.
Pearson Education—Japan, *Tokyo*
Pearson Education Malaysia, Pte. Ltd.

CONTENTS

PREFACE

This is the eleventh edition of this laboratory manual that was first published in 1964. During the intervening 37 years, our knowledge of the forces that shape our planet has grown exponentially. The revolutionary theory of plate tectonics is firmly established and provides a framework for learning about Earth's dynamics and the interrelationships between moving tectonic plates, mountain building, the origin of ocean basins, and the evolution of continents.

New technology permits us to see Earth from space, image the ocean floor, and measure directly the motion of tectonic plates. We can see in one synoptic view the surface of an entire continent and map ocean currents, temperatures, and vegetation from satellites orbiting in space. During no other period has there been so much exploration and development of new knowledge about Earth. For this reason, we have revised this manual in an attempt to incorporate the new theories and discoveries.

The objectives we set forth in the first edition still stand:

1. To give students experience in examining geologic data and formulating hypotheses to explain observed facts.

2. To provide an opportunity to continue laboratory-type work outside of class so students can prepare adequately for lab sessions and review work independently.

3. To give laboratory instructors maximum latitude in their instruction by providing abundant material from which they can select for their own specific objectives.

NEW TO THIS EDITION

Every exercise has been carefully updated and checked for accuracy. New questions and problems were developed, where appropriate, but previous problems that have "stood the test of time" have been retained. Examples of changes include extensive use of new computer-generated shaded relief maps, new Landsat images and aerial photographs, and various new remote-sensing images. We have, however, retained many of the classic maps and aerial photographs that have served effectively as standard exercises for many years. We have also made some significant changes in organization so that review and discussion of the basic principles establish a solid background for problem solving.

Rocks and Minerals

Most of the photographs of rocks and minerals are the same as those used in the previous edition, but some have been replaced where photographs of better specimens were obtainable. This material is intended as reference material for comparison with laboratory specimens. Photographs, of course, can never replace study of actual hand specimens, but they are useful as a guide and reference in the study of physical properties of minerals and textures of rocks.

Maps, Aerial Photographs, and Remote Sensing Images

Maps, aerial photographs, and various types of remote sensing images are the fundamental tools of geologic research and are naturally the basic materials in laboratory work of physical geology. The advances in geology during the last several decades have brought about profound changes in the making of maps. As a result of the space program, we now have sophisticated satellite imagery of Earth's surface and radar images that can "see through" clouds and vegetation cover. With remote sensing, we can also see the detailed landscape of the ocean floor and recognize features as small as a submerged boat. We can observe what was once unseen, and we can view the surface features of our planet from exciting new perspectives. Landsat images of Earth can be enhanced by the computer—enlarged, manipulated in tone and color, and even reconstructed to produce stereoscopic images. In addition, most of the United States has been photographed with high-altitude infrared photography, and radar images have been made of large areas of North America. These exciting new images are the basic data for many of the exercises in the tenth edition.

Perhaps the most significant recent advance in mapmaking has been the development of digital shaded relief maps that show the surface features of Earth in relief and remarkable detail. Since graphic representations of Earth's surface features are fundamental to studying and understanding geology, we introduce students to these new maps and images and involve them in the interpretation of the geologic processes revealed by each.

Structural Geology

We have retained the series of diagrams illustrating the major structural features and their outcrop patterns. Portions of the geologic map of the United States have been retained from previous editions, and new maps, radar images, and computer-enhanced Landsat images have been added in the exercise section.

Plate Tectonics

The theory of plate tectonics has influenced every aspect of geology and has focused our attention on the global aspects of the science. To give students experience in analyzing geologic features on a global scale, we include a large physiographic map of Earth that serves as a basis for exercises in plate tectonics, major structural features of the continents, and geology of the ocean floor.

Seismology and Earth's Interior

We have retained the exercise in seismology that introduces students to the way geologists study Earth's interior. Problems in seismic stratigraphy and the study of shallow geologic structures give students a chance to work with seismic records. This exercise also includes studies of P and S wave shadow zones so students can see how scientists determine the nature of Earth's deep internal structure. In addition, we have added an exercise utilizing interferometric maps enabling the student to study surface deformation associated with earthquakes.

Planetary Geology

The exploration of the planets has added yet another dimension to the study of Earth because it permits us to compare and contrast the geologic systems of other planetary bodies with those of our planet. We have revised this section to emphasize detailed examples of the new images of Mars and Europa, with resolutions of several tens of meters, permitting us to see features never seen before. The study of other planetary bodies serves as contrasts to the geologic systems on Earth.

ACKNOWLEDGMENTS

We are grateful to our colleagues for their many constructive criticisms, comments, and suggestions. The wide diversity in scope and content of the laboratory experience in physical geology is apparent. We have altered exercises when the consensus was clear to do so, but have relied on our own judgment when opinions varied.

We also thank Wards Natural History Establishment, Rochester, New York, for letting us use many specimens to make photographs of rocks and minerals, and the Photo Mechanix Studios, Bloomington, Minnesota, for their excellent work in photographing rock and mineral specimens.

We are especially grateful to Geo-Science Resources for the excellent photomicrographs used in the rocks and minerals section. Extra effort by the Geo-Science Resources staff to obtain the best specimens available combined with their high quality photography provides students with a unique visual aid to the understanding of rock and mineral composition.

The U.S. Geological Survey topographic maps that appear in this book are made from the original plates created by the U.S. Geological Survey. These maps are specially edited and designed for teaching purposes and are not intended to duplicate the complete maps. We are grateful to the U.S. Geological Survey for permission to reproduce these topographic maps, sections of geologic maps, and aerial photographs.

We give special thanks to Michael J. Oimoen of Eros Data Center, Sioux Falls, South Dakota, who helped provide the most recent and detailed digital shaded relief maps from the U.S. Geological Survey. We also thank the Agricultural Stabilization and Conservation Services of the U.S. Department of Agriculture for the use of aerial photographs from their files. The aerial photographs on pages 69 (top), 113 (bottom), 136 (top), 146, and 177 (top) © Her Majesty the Queen in Right of Canada, reproduced from the collection of the National Air Photo Library with permission of Natural Resources Canada.

We sincerely thank Dale Claflin, who spent many hours preparing illustrations for previous editions, and Rick Showalter, who prepared new line drawings for this edition.

The unique computer-generated shaded relief maps made by Kenneth Perry of Chalk Butte Inc. are especially appreciated.

MINERALS AND CRYSTAL GROWTH

OBJECTIVE

To gain an understanding of the crystalline structure of minerals by observing the process of crystal growth.

MAIN CONCEPT

Minerals grow by crystallization, a process by which atoms are arranged in a specific geometric pattern.

SUPPORTING IDEAS

1. Crystal form is an expression of a mineral's atomic structure.
2. All crystals of a given mineral possess the same properties of symmetry, even though the crystals may differ in size and shape.
3. Similar pairs of crystal faces of a given mineral always meet at the same angle (law of constancy of interfacial angles).

THE NATURE OF MINERALS

A mineral is composed of elements or groups of elements that unite in nature to form an inorganic crystalline solid. Every crystalline substance has a definite internal structure in which the atoms occur in specific proportions and are arranged in an orderly geometric pattern. This systematic arrangement of atoms is one of the most significant aspects of a mineral. It exists throughout the entire specimen, and if crystallization occurs under ideal conditions, the arrangement will be expressed in perfect crystal faces (Figure 1.1).

For example, moisture in the air may freeze and develop into ice crystals. *The form of the solid ice crystals is an external expression of the orderly arrangement of the water molecules.* Although each individual crystal of ice is different in size and shape, all possess the same properties of symmetry. Similarly, a solution rich in sodium (Na) and chlorine (Cl) ions will, on evaporation, develop into crystals of halite (NaCl), or common salt. All halite crystals will have the same internal structure and the same properties of symmetry in their crystal form.

Not all mineral crystals have perfect faces or perfect crystal form because of restrictions in the environment in which they crystallize. Minerals with poor or imperfect crystal faces do still possess, however, a systematic internal structure, as revealed by properties such as cleavage and the symmetrical patterns produced when a thin beam of X-rays is passed through them. In contrast, the atoms in a liquid or gas have little or no systematic structure and no definite form. Substances that become solid without crystallizing are *amorphous*—that is, without a definite internal structure. Glass, for example, is rigid and solid, but the atoms of glass are not arranged in a systematic pattern. Therefore, physicists regard glass as an extremely viscous liquid.

A significant feature that expresses the internal structure of crystals is the constancy of their interfacial angles. The angles between similar crystal faces of a specific mineral will be the same, even though the size of the crystals may vary. This unique feature is of prime importance in differentiating minerals that are similar in other respects. Numerous precise measurements of the angles between crystal faces have demonstrated repeatedly that although crystals of the same substance may exhibit different overall shapes, the angles between corresponding crystal faces are identical. This phenomenon is known as the *law of constancy of interfacial angles.*

GROWTH OF MINERALS

Minerals grow as matter changes from a gaseous or liquid state to a solid state, or when one solid recrystallizes to form another. Growth is accomplished by a process known as *crystallization*—the arrangement of atoms into a geometric structure, held together by atomic bonds. Growth occurs by addition of atoms to a crystal face. Crystallization (mineral growth) is possible because the outer layers of atoms on a crystal are never complete and more atoms can be added indefinitely. An environment suitable for crystals to grow includes (1) proper concentration of the kinds of atoms or ions required for a particular mineral, and (2) proper temperature and pressure. A single crystal may range in size from submicroscopic (Figure 1.1F) to more than 10 meters long. Theoretically, a crystal can grow indefinitely if the proper chemical and physical conditions exist.

In this exercise, you will grow crystals under various conditions and observe this fascinating process in action.

A. Crystals of Wulfenite, Sonora, Mexico (8.7 cm high). *(Jeffrey A. Scovil)*

B. Crystals of Rhodochrosite, Sweet Home MIne, Colorado (2.5 cm wide). *(Jeffrey A. Scovil)*

C. Crystals of Elbaite and Topaz, Pech, Afghanistan (6.5 cm high). *(Jeffrey A. Scovil)*

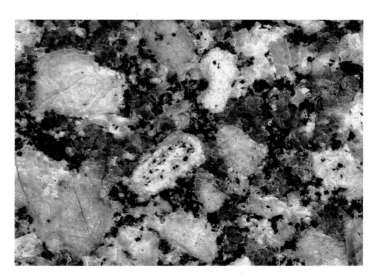

D. Crystals in Granite, Rockville, Minnesota (actual size)

E. Crystals in Gabbro, as seen under the microscope (x30)

F. Crystals of Clay Minerals in voids between sand grains, as seen with the electron microscope (x2000)

FIGURE 1.1
Examples of Crystals

A. GROWTH OF CRYSTALS FROM SOLUTION

Many minerals are precipitated from aqueous solutions by evaporation at atmospheric temperature and pressure. Crystal growth of such minerals can be observed in the laboratory by evaporating prepared concentrated solutions.

Your instructor will provide concentrated solutions of sodium chloride, sodium nitrate, potassium aluminum sulfate (alum), and copper acetate.

Place a drop of each solution on a slide. Label each solution on a piece of tape placed at the end of the slide. As the water evaporates, crystals of each compound will appear, and as evaporation continues, the crystals will grow larger. Observe the slide through a microscope or magnifying glass at several different times during the process of crystal growth. (While you are waiting for sufficient evaporation to occur to form crystals, you may begin the next part of this exercise concerning the growth of crystals from a melt.)

1. Make a sketch showing the outline of two crystals from each compound.

2. Which compound has a cubic (box-like) form?

3. Why is it that not all of the crystals are perfect cubes?

4. Do the crystals of each compound satisfy the law of constancy of interfacial angles?

5. Does each compound have a unique crystal form?

6. Can minerals be identified from crystal form alone?

B. GROWTH OF CRYSTALS FROM A MELT

Most minerals that originate from a melt (or magma) crystallize at temperatures well above the boiling point of water, so directly observing their crystal growth is difficult. Thymol, an organic chemical, crystallizes near room temperature; although it differs from a magma, certain principles of crystallization demonstrated in a thymol melt pertain generally to crystallization of a magma. Thymol is not poisonous but should be handled with forceps, for it can irritate the skin and eyes.

In performing the experiment, remember to do the following:

1. Observe crystal growth and distinguish a single crystal from an aggregate of crystals.

2. Study the law of constancy of interfacial angles by observing crystals growing from a melt solution.

3. Discover, for yourself, some of the factors that govern crystal growth.

Slow Cooling without "Seed" Crystals

Place a petri dish containing a small amount of crystalline thymol on a hot plate, adjusted to low heat, until all of the crystals are melted. Let the melt continue to heat for 1 or 2 minutes more, then remove the dish from the hot plate and set it aside to cool slowly where it will not be disturbed. You will examine this melt near the end of the lab.

Slow Cooling with "Seed" Crystals

Repeat the procedure just described, but transfer the petri dish to the stage of the microscope *as soon as the thymol is melted*. Add four or five small grains, or "seed" crystals, to the melt, and observe the melt under the microscope or with a magnifying lens. If the seed crystals melt, wait for a few moments and add more of them. As the melt cools, observe the crystals beginning to grow.

1. Sketch the shape of one or two single crystals and describe briefly the manner of crystal growth. (Consider such factors as internal expansion versus external accretion, rate of growth, direction of growth, angles between crystal faces, the effect of limited space on crystal shape, the difference between a single crystal and an aggregate of crystals, and the texture produced by interlocking crystals.)

2. How do the growth lines of a crystal compare with the outline of a single crystal in a solid aggregate?

3. What role do the seed crystals play in initiating crystal growth?

4. When less than half of the melt remains, you may notice the formation of some spontaneous nucleating crystals. Study these as they grow. Make a sketch of the final crystalline aggregate. Individual crystals can be recognized by their characteristic growth lines and by how they reflect light. The aggregate of crystals produced in this experiment is similar to that found in many common igneous rocks. Both are composed of many interlocking crystals originating from a melt.

5. Examine several specimens of granite supplied by your instructor. Discriminate between the individual crystals, and sketch the texture of the granite. How does the texture of an aggregate of thymol crystals compare with the texture of the minerals in a granite?

Rapid Cooling

Place a petri dish containing a small amount of thymol on a hot plate until only a few crystals remain. Transfer the dish to the top of an ice cube, and let the melt cool there for a fraction of a minute. As the melt cools rapidly, observe the crystals beginning to grow. Almost immediately, several independent, spontaneous nucleating centers of crystallization will develop. Quickly transfer the petri dish to the stage of a microscope and observe the nature of the crystal growth. Repeat the experiment as often as time permits, so that you are sure of your general observations and conclusions.

Comparison of Crystals Produced by Slow Cooling and Rapid Cooling

Examine the thymol in the dish prepared for the first part of this exercise concerned with slow cooling without seed crystals.

1. What effect does the rate of cooling have on crystal size?

C. INTERPRETING THE RESULTS OF MINERAL GROWTH

With some experience growing your own crystals, you should now be able to observe various specimens of minerals, such as the ones in the photographs below, and draw certain conclusions about the way in which they form.

1. Did the crystals in (A) form in a cavity or cave or did they form a lava flow? Cite evidence to defend your answer.

2. Two different minerals crystallized to form the specimen in (B). Did they grow at the same time, or are two distinct periods of crystallization represented?

3. The specimen in (C) is an igneous rock (the minerals crystallized from a melt). Did all the minerals grow at the same time or did some form earlier than others?

4. (D) is a view of a thin slice of rock through a microscope. How many periods of mineral growth can you recognize? Cite evidence for your answer.

5. The specimen in (E) is a cluster of salt crystals (the mineral halite) that grew on a wire fence near the shore of Great Salt Lake. Briefly explain why the crystals grew.

A

(Jeffrey A. Scovil)

B

(Jeffrey A. Scovil)

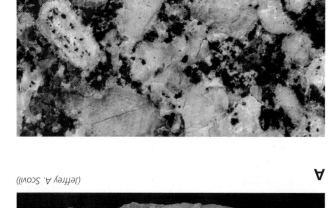

C

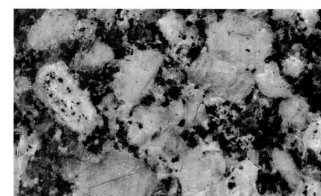

D

E

(Laurel Casjens, Utah Museum of Natural History, University of Utah)

EXERCISE 2

MINERAL IDENTIFICATION

■ OBJECTIVE

To become familiar with the common physical properties of minerals and to be able to use these properties in mineral identification.

■ MAIN CONCEPT

Each mineral species is characterized by a specific (atomic) internal structure and by a specific chemical composition that varies only within certain limits. As a result, all speci-mens of a given mineral will possess certain diagnostic physical properties.

■ SUPPORTING IDEAS

1. The more important physical properties of minerals are (a) crystal form, (b) cleavage, (c) hardness, (d) density, (e) streak, and (f) luster.
2. Other physical properties such as color, tenacity, reaction with hydrochloric acid, and magnetism are important diag-nostic properties of some minerals.
3. The physical properties of a mineral can be used to iden-tify mineral specimens.
4. The most important rock-forming minerals are quartz, feldspar, olivine, pyroxene, amphibole, mica, calcite, and clay minerals.

PHYSICAL PROPERTIES OF MINERALS

Crystal Form

When a crystal is allowed to grow in an unrestricted environment, it will develop natural crystal faces that pro-duce a perfect geometric pattern. The shape of such a crystal is a reflection of its internal structure and can be used to identify many mineral species. Remem-ber, however, that two or more differ-ent minerals may have essentially the same internal structure and may develop similar crystals. Some of the common minerals in which crystal form is especially diagnostic are quartz, halite, garnet, fluorite, pyrite, galena, amphibole, and pyroxene.

Cleavage

The forces that hold the atoms together in a crystalline structure are not necessar-ily equal in all directions. If definite planes of weakness exist, the mineral will cleave, or break, along the planes of weakness much more easily than in other directions. The surface along which the break develops is referred to as the *cleavage plane*, and the orientation of the plane is the *cleavage direction*.

Perfect cleavage is recognized eas-ily, for it develops a characteristically smooth, even surface that reflects light like a mirror (Figure 2.1A). Good exam-ples of cleavage are shown in Figures 2.7B and D, and 2.8D, E, and J. Cleav-age planes can occur in small segments arranged in a step-like manner, how-ever, and the surface of the specimen may appear at first to have an irregular fracture. If the specimen is rotated in front of a light, these small, parallel cleavage planes will reflect light just as a large, smooth cleavage surface does (Figure 2.1B). In contrast, an uneven fracture will not concentrate light in any particular direction (Figure 2.1C).

Keep in mind the distinction between cleavage planes and crystal faces. *Cleavage results from planes of weakness within the crystal structure along which the crystal breaks. Crystal faces reflect the geometry of the atomic structure.* Also remember that both are properties of single crystals. In a crys-talline aggregate, each crystal will break along its own cleavage planes, if any exist, but the aggregate does not pos-sess a specific cleavage.

Crystals may have one, two, three, four, or six cleavage directions, as illus-trated in Figure 2.2. Some minerals, such as gypsum, cleave in multiple

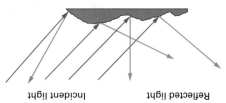

A. Reflected Light from Smooth Cleavage Surface

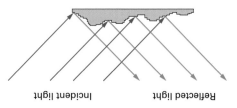

B. Reflected Light from Stepped Cleavage Surfaces

C. Reflected Light from Fracture

FIGURE 2.1
Reflection of Light from Cleavage and Fracture Surfaces

FIGURE 2.2
Crystal Cleavage

A. Cleavage in One Direction. Example: muscovite.

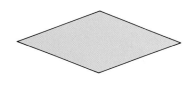

B. Cleavage in Two Directions at Right Angles. Example: feldspar.

C. Cleavage in Two Directions Not at Right Angles. Example: amphibole.

D. Cleavage in Three Directions at Right Angles. Example: halite.

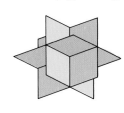

E. Cleavage in Three Directions Not at Right Angles. Example: calcite.

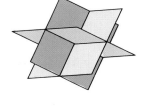

F. Cleavage in Four Directions. Example: fluorite.

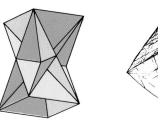

G. Cleavage in Six Directions. Example: sphalerite.

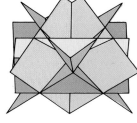
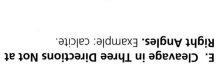

directions but possess better cleavage in one direction than in another.

In some minerals, the crystal structure is so well knit that there is no tendency to break along one plane in preference to another. Such minerals do not possess cleavage, but break or fracture in an irregular manner. Common fracture types are irregular, conchoidal, and fibrous.

Hardness

Hardness is a measure of a mineral's ability to resist abrasion. Like cleavage and fracture, this property is also related to crystal structure and to interatomic bond strength. Hardness is a relative property and was one of the first properties used in the field to identify unknown minerals. If two minerals are rubbed together, the harder one scratches the softer. Over a century ago, the German mineralogist Friedrich Mohs developed a hardness scale by assigning an arbitrary relative number to ten common minerals. Diamond, the hardest mineral known, was placed at the top of his hardness scale and

assigned the number 10. Softer minerals were ranked in descending order, perhaps without a realization of the unequal degree of hardness between the various ranks. More scientific hardness tests have been devised since, in which the absolute hardness values can be measured. A comparison of the Mohs scale and an absolute hardness scale is shown in Figure 2.3. Among the softer minerals there is little difference in quantitative hardness, but rather large steps occur between quartz, topaz, and corundum, and above all between corundum and diamond.

The Mohs hardness scale and the hardness of some common everyday objects are shown in Figure 2.3. These objects can be used effectively to test the hardness of many unknown minerals. For example, calcite will scratch a fingernail, gypsum, and talc, but it will not scratch a knife blade, fluorite, or quartz. The Mohs scale is not exact, but its convenience more than makes up for its lack of precision. It is used often as a simple preliminary test of a mineral's hardness property.

Hardness is generally a reliable diagnostic physical property of a mineral, but variations in composition may render some mineral specimens harder or softer than normal. Moreover, because weathering can affect hardness, making tests on fresh surfaces is important.

Density

Density is the mass per unit volume of a substance, commonly expressed in grams per cubic centimeter. It is one of a mineral's most constant physical properties. For purposes of general laboratory work and field work, however, you can estimate density with surprising accuracy simply by lifting a mineral specimen in your hand and making an estimate. Galena, with a density of approximately 7.5, and pyrite, with a density of about 5, are typical of minerals with high density. They are distinguished easily from common rock-forming minerals such as quartz, feldspar, and calcite, each of which has a density of between 2.6 and 2.8.

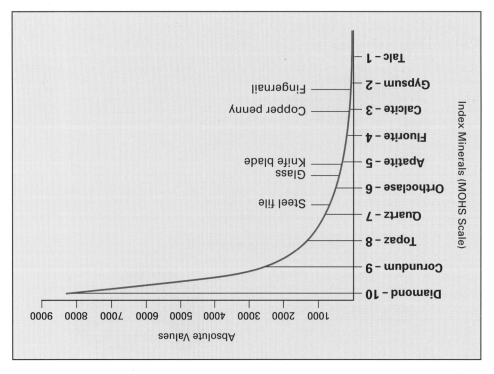

FIGURE 2.3
Hardness Scale. Mohs hardness scale compared with an absolute hardness scale and some common everyday objects.

Color

Color is one of the most obvious physical properties, and for some minerals—such as galena (gray), azurite (blue), and olivine (green)—it is diagnostic. Other minerals are found in differing hues, which may be caused by variations in composition or inclusion of impurities. Quartz, for example, ranges through a spectrum of colorless clear crystals to purple, red, white, and jet black. Although color may be diagnostic for a few minerals, in others it is almost without diagnostic significance. *Color should be considered in mineral identification but should never be used as the major identifying characteristic.*

Streak

When a mineral is powdered, it usually exhibits a much more diagnostic color than when it occurs in large pieces. The color of the powdered mineral is referred to as *streak*. In the laboratory, streak is obtained by rubbing the mineral vigorously on an unglazed porcelain plate. Commonly, a mineral's streak will be different from the color seen in the hand specimen; thus, do not anticipate the streak color by a simple visual examination of a mineral fragment.

Most minerals with a nonmetallic luster have a white or pastel streak. For this reason, streak is not very useful in distinguishing nonmetallic minerals. Minerals with a hardness greater than that of porcelain will scratch the plate and will not produce streak.

Luster

The terms *metallic* and *nonmetallic* describe broadly the basic types of *luster*, or the appearance of light reflected from a mineral. Pyrite and galena have a metallic luster on unweathered surfaces. A variety of nonmetallic lusters can be distinguished, the most important of which are described as vitreous (glassy), pearly, resinous, silky, and greasy. To become acquainted with these, study examples of each luster type from specimens provided by your instructor. Some minerals, because of their characteristically weathered or porous nature, have a distinctive dull or earthy luster.

Tenacity

The manner in which a mineral resists breakage is its *tenacity*. This property is described by the following terms:

1. Brittle—crushes to angular fragments. Example: quartz.

2. Malleable—can be modified in shape without breaking and can be flattened to a thin sheet. Example: native copper.

3. Sectile—can be cut with a knife into thin shavings. Example: talc.

4. Flexible—will bend but does not regain its original shape when force is released. Example: gypsum.

5. Elastic—will bend and regain its original shape when force is released. Example: muscovite or biotite.

Optical Properties

Minerals have a number of important optical properties, most of which require a polarizing microscope for effective study. However, the general way in which a mineral transmits light can be observed in hand specimens and is useful in identifying some minerals. Optical properties are described as follows:

1. Transparent—objects are visible when viewed through the mineral. Example: some species of quartz, calcite, and biotite.

2. Translucent—light, but not an image, is transmitted through the mineral. Example: some varieties of gypsum.

3. Opaque—no light is transmitted, even on the thinnest specimen edges. Example: magnetite or pyrite.

Other Properties

Reaction to Hydrochloric Acid. Calcite ($CaCO_3$) is one of the most common minerals on Earth's surface, and when treated with dilute hydrochloric acid (HCl), it will effervesce, or bubble, vigorously. This simple chemical test is very diagnostic and can be used to distinguish calcite from most of the other common minerals. Dolomite ($CaMg[CO_3]_2$), a mineral similar to calcite, will react with cold, dilute hydrochloric acid, but only if the specimen is powdered.

Double Refraction. A property of the transparent variety of calcite, whereby light passing through the crystal or cleavage fragment is split into two rays, is known as double refraction. An object viewed through the mineral therefore shows a double image.

Magnetism. Magnetite is one of the few minerals to show obvious magnetic attraction. Some other minerals do have distinctive magnetic properties that, with proper equipment, can be used for mineral separation, but magnetite is the only common mineral actually attracted by a small magnet.

Taste. The salty taste of halite is a definite and unmistakable property of that mineral. Other evaporite minerals usually have a bitter taste, but they are not exceedingly common.

Ionic Substitution

Various ions can substitute for each other in the structure of some minerals. When this occurs, there is a chemical change in the mineral but no change in its structure. Several factors determine the suitability of one ion to substitute for another; the most important are the size and electrical charge of the ions in question. As a result, the composition of a mineral may vary with specific limits. This may result in variations within physical properties such as hardness and color.

Mineral Identification Chart

The chart in Figure 2.4 shows common rock-forming minerals grouped according to the categories of (1) luster and color, (2) hardness or streak, and (3) cleavage or fracture. For each mineral, additional physical characteristics are briefly described; the most diagnostic characteristics appear in bold type. Photographs of some common minerals appear in Figures 2.5 through 2.9, together with brief captions.

		Good cleavage in 2 directions at approximately 90°; commonly light to dark pink, pearly to vitreous luster; H = 6–6.5; D = 2.5.	POTASSIUM FELDSPAR: $KAlSi_3O_8$
Harder than glass	**Cleavage prominent**	Good cleavage in 2 directions at approximately 90°; white to gray; striations on some cleavage planes.	PLAGIOCLASE FELSPAR: $NaAlSi_3O_8$ to $CaAl_2Si_2O_8$
	Cleavage absent	Conchoidal fracture; H = 7; D= 2.65; transparent to translucent; vitreous luster;6-sided prismatic crystals terminated by 6-sided triangular faces in well-developed crystals; vitreous to waxy; colors range from milky white, rose pink, and violet to smoky gray.	QUARTZ: SiO_2 (silica) Varieties milky; smoky; rose; amethyst
		Concoidal fracture; H = 7; variable color; translucent to opaque; dull or clouded luster; colors range widely from white, gray, and red to black.	CRYPTOCRYSTALLINE QUARTZ SiO_2 Varieties agate; flint; chert; jasper; opal

NON-METALLIC LUSTER LIGHT COLOR

		Perfect cubic cleavage; salty taste; colorless to white; soluble in water; H = 2–2.5; D = 2.	HALITE: NaCL
		Perfect cleavage in 1 direction; poor in 2 others; H = 2; white; transparent; nonelastic; D = 2.3. Varieties: Selenite: colorless, transparent Alabaster: aggregates of small crystals Satin spar: fibrous, silky luster	GYPSUM: $CaSo_4 • 2H_2O$
	Cleavage prominent	Perfect cleavage in 3 directions at approximately 75°; effervesces in HCL; H = 3; colorless, white, or pale yellow—rarely gray or blue; transparent to opaque; D = 2.7.	CALCITE: $CaCO_3$ (fine-grained crystalline aggregates form limestone and marble)
Softer than glass		Three directions of cleavage as in calcite; effervesces in HCl only if powdered; H = 3.5–4; D = 2.8; color variable but commonly white or pink; rhomb-faced crystals.	DOLOMITE: $CaMg(CO_3)_2$
		Good cleavage in 4 directions; H =4; D = 3; colorless, yellow, blue, green, or violet; transparent to transluent; cubic crystals.	FLUORITE: CaF_2
		Perfect cleavage in 1 direction, producing thin, elastic sheets; H = 2–3; D = 2.8; transparent and colorless in thin sheets.	MUSCOVITE: $KAl_3AlSi_3O_{10}(OH)_2$
		Green to white; soapy feel; pearly luster; H = 1; D = 2.8; foliated or compact masses; one direction of cleavage forms thin scales and shreds.	TALC: $Mg_3Si_4O_{10}(OH)_2$
	Cleavage absent	White to red; earthy masses; crystals so small no cleavage is visible; soft, H = 1.2; becomes plastic when moistened; earthy odor.	KAOLINATE: $Al_4Si_4O_{10}(OH)_8$

FIGURE 2.4
Mineral Classification Chart

NOTE: The most diagnostic properties for each mineral are indicated by bold type.

NON-METALLIC LUSTER, DARK COLOR	**Harder than glass**	**Cleavage prominent**	**Cleavage—2 directions at nearly 90°;** dark green to black; short, prismatic 8-sided crystals; H = 6; D = 3.5.	PYROXENE GROUP: Complex Ca, Mg, Fe, Al silicates
			Cleavage—2 directions at approximately 60° and 120°; dark green to black or brown; long, prismatic 6-sided crystals; H = 6; D = 3.35.	AMPHIBOLE GROUP: Complex Na, Ca, Mg, Fe, Al silicates
			Good cleavage in 2 directions at approximately 90°; gray to blue-gray; striations on some cleavage planes.	PLAGIOCLASE FELDSPAR: $NaAlSi_3O_8$ to $CaAl_2Si_2O_8$
		Cleavage absent	**Various shades of green; sometimes yellowish; commonly occurs in aggregates of small glassy grains;** transparent to translucent; glassy luster; H = 6.5–7; D = 3.5–4.5.	OLIVINE: $(Fe, Mg)_2SiO_4$
			Red, brown, or yellow; glassy luster; conchoidal fracture resembles poor cleavage; commonly occurs in well-formed 12-sided crystals; H = 7–7.5; D = 3.5–4.5.	GARNET GROUP: Fe, Mg, Ca, Al silicates
			Conchoidal fracture; H = 7; gray to gray-black; vitreous luster.	QUARTZ: SiO_2
	Softer than glass	**Cleavage prominent**	**Brown to black; 1 perfect cleavage;** flexible and elastic when in thin sheets; H = 2.5–3; D = 3–3.5.	BIOTITE: $K(Mg, Fe)_3AlSi_3O_{10}(OH)_2$
			Green to very dark green; 1 cleavage direction; commonly occurs in foliated or scaly masses; nonelastic plates; H = 2–2.5; D = 2.5–3.5.	CHLORITE: Hydrous Mg, Fe, Al silicate
			Yellowish brown; resinous luster; cleavage in 6 directions; yellowish brown or nearly white streaks; H = 3.5–4; D = 4.	SPHALERITE: ZnS
			Four perfect cleavage directions; H = 4; D = 3; green through deep purple; transparent to translucent; cubic crystals.	FLUORITE: CaF_2
		Cleavage absent	**Red, earthy appearance;** red streak; H = 1.5.	HEMATITE: Fe_2O_3 (earth variety)
			Yellowish-brown streak: yellowish brown to dark brown; commonly in compacted earth masses; H = 1.5.	LIMONITE: $Fe_2O_3 \cdot H_2O$

METALLIC LUSTER	**Black, gray, greenish black streak**	**Perfect cubic cleavage;** heavy, D = 7.6; H = 2.5; **silver-gray color,** bright metallic luster.	*GALENA: PbS*
		Magnetic; black to dark gray; D = 5.2; H = 6; commonly occurs in granular masses; single crystals are octahedral.	*MAGNETITE: Fe_3O_4*
		Steel gray; soft; smudges fingers and marks paper; greasy feel; H = 1; D = 2; luster may be dull.	*GRAPHITE: C*
		Golden yellow; may tarnish purple; H = 4; D = 4.3; streak is greenish black.	*CHALCOPYRITE: $CuFeS_2$*
		Brass yellow; cubic crystals; striated facet common; common in granular aggregates; H = 6–6.5; D = 5; uneven fracture.	*PYRITE: FeS_2*
	Brown to reddish-brown streak	**Steel gray; red to red-brown streak;** black to dark brown; granular, fibrous, or micaceous; single crystals are thick plates; H = 5–6; D = 5; uneven fracture.	*HEMATITE: Fe_2O_3*
	Yellow, brown streak	**Yellow, brown, or black;** hard, structureless or radial fibrous masses; H = 5–5.5; D = 3.5–4; yellow-brown streak.	*LIMONITE: $Fe_2O_3 \cdot H_2O$*

Varieties of Quartz

Quartz is one of the most common minerals you are likely to encounter because it is a major constituent in many igneous, sedimentary, and metamorphic rocks. If allowed to grow in an unrestricted environment, quartz will form well-developed hexagonal (six-sided) crystals that terminate in a pyramid. The crystal form of quartz is thus a distinctive property. In most rocks, however, the crystal form of quartz grains is not expressed. In igneous rocks and in veins, quartz is one of the last minerals to form, so it fills the available space between other minerals, and the quartz grains thus have irregular shapes. In sedimentary rocks, quartz grains are rounded by the abrasion they have undergone during transportation by wind and water. In metamorphic rocks, quartz grains and other mineral grains are generally deformed and recrystallized. The quartz may therefore appear in a variety of sizes and shapes.

The color, crystal size, and general appearance of quartz vary greatly, so a large number of varieties of quartz have been named. These can be classified in two major groups: (1) macrocrystalline and (2) cryptocrystalline. Further subdivision in each group can be made on the basis of color or some other special feature. However, regardless of such features as color, crystal size, crystal shape, and mode of origin, all quartz is characterized by the following properties: (1) hardness of 7 (it will scratch glass and steel), (2) conchoidal fracture, and (3) glassy luster.

Macrocrystalline quartz is composed of individual grains, or crystals, that can be seen with the naked eye or with low-power magnification. Varieties are distinguished mainly on the basis of color—the major varieties being clear, rose, milky, smoky, and amethyst. Cryptocrystalline quartz is composed of aggregates of innumerable submicroscopic crystals. Color variations give rise to many varieties, such as chert, flint, agate, jasper, and opal (See Figure 2.5J–O).

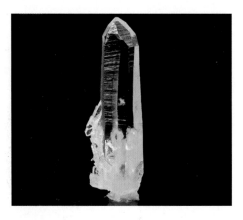

A. Quartz Crystals often develop in groups of slender, six-sided prisms that terminate in a pyramid. Striations are common across the prism faces. Crystals are usually milky, colorless, or transparent.

B. Quartz Crystals grow in a variety of sizes, shapes, and colors. In every quartz crystal—regardless of where, when, or how it grew—the angles between corresponding crystal faces are identical.

C. Amethyst is a transparent quartz crystal in purple or violet hues that usually develops as a secondary mineral in veins and cavities. The color is due to small inclusions of ferric iron. *(Jeffrey A. Scovil)*

D. Milky Quartz occurs in veins and is commonly massive, without well-developed crystal faces. Minute fluid inclusions produce the white color.

E. Smoky Quartz is brown, gray, or black and is characteristic of intrusive igneous rocks. It is transparent to translucent in luster. Radiation from radioactive minerals may be responsible for the color of smoky quartz. *(Jeffrey A. Scovil)*

F. Rose Quartz owes its color to small amounts of titanium. Rose quartz is generally coarsely crystalline, transparent to translucent, and rarely forms well-developed crystals.

FIGURE 2.5
Varieties of Quartz

G. Quartz in Granite. Quartz is one of the last minerals to form in granite and thus fills the spaces between the crystals of other minerals formed at higher temperatures. Quartz in granite is characterized by a glassy luster and conchoidal fracture and is usually smoky.

H. Quartz in Sandstone. Most sand grains are quartz crystals that have been rounded and abraded during transportation by wind or water. The surface of some grains may be pitted or frosted. Broken grains show a glassy luster and conchoidal fracture.

I. Quartz in Quartzite. Heat and pressure may fuse sand grains to form a dense, massive rock called quartzite. Quartzite is resistant to weathering and abrasion, so that quartzite pebbles are common constituents of many gravels.

J. Chert is a light-colored variety of cryptocrystalline quartz. It occurs commonly as nodules in limestone.

K. Flint is similar to chert but is characterized by its dark color. Both break with a conchoidal fracture.

L. Agate is a variety of cryptocrystalline quartz in which colorful banding occurs due to successive periods of deposition. *(Jeffrey A. Scovil)*

M. Jasper is a variety of cryptocrystalline quartz with a distinctive red color produced by minute inclusions of hematite.

N. Opal is a variety of cryptocrystalline quartz that contains considerable amounts of water. It is distinguished from other varieties by its waxy luster. It is commonly capable of refracting light and then reflecting it in a play of colors. *(Jeffrey A. Scovil)*

O. Petrified Wood. Cryptocrystalline quartz commonly replaces organic material in the process of fossilization. In petrified wood, many delicate structures are preserved because of the small crystal size.

THE FELDSPAR GROUP

Feldspar is the most abundant mineral in Earth's crust. Like quartz, it is a basic constituent of many igneous, sedimentary, and metamorphic rocks. All feldspars are aluminum silicates containing either potassium, sodium, or calcium, and all are related closely in form and physical properties. Two main subgroups are recognized: (1) the *plagioclase feldspars,* which range in composition from $CaAl_2Si_2O_8$ to $NaAlSi_3O_8$, and (2) the *potassium feldspars,* which have a composition of $KAlSi_3O_8$. Varieties of potassium feldspar occur because of differences in the arrangement of Al and Si ions in the crystal structure. The varieties of feldspar can be distinguished best by chemical analysis or by optical or X-ray measurements of the crystal structure. Color in feldspars is quite variable and is due to small amounts of iron or magnesium and to certain aspects of the crystal structure. Color is *not* diagnostic of a particular variety of feldspar. *The most important physical properties of the feldspars are (1) two-directional cleavage at approximately right angles, (2) hardness of 6, and (3) pearly luster.*

A. Sodium Plagioclase (Albite). Sodium plagioclase feldspar is commonly light colored and has a porcelain luster. The two-directional cleavage, characteristic of all feldspars, is well expressed in most specimens.

B. Calcium Plagioclase (Labradorite). Striations on certain cleavage surfaces are diagnostic of calcium plagioclase and distinguish this group from the potassium feldspars. Ca-plagioclase is dark colored and shows iridescence.

C. Potassium Feldspar. Although color is generally not diagnostic, most pink and green feldspars belong to the potassium subgroup. Well-developed cleavage in two directions is similar to that in plagioclase feldspar.

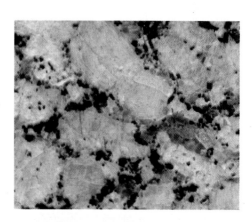

D. Feldspars in Granite. Potassium feldspar is a major constituent of granite and is commonly expressed by rectangular pink crystals. Sodium feldspar in granite is characteristically white. Both may occur in the same rock specimen. In this specimen, the pink grains are potassium feldspar, the milky-white grains are Na-plagioclase, and the smoky, glassy grains are quartz.

E. Feldspar in Sandstone. Weathering and erosion of a granite will disaggregate the feldspar and quartz. Feldspars break down rapidly by weathering to form clay minerals and are therefore not a major constituent of most sediments. Angular grains of feldspar such as those shown in this specimen, however, are found in some sandstones deposited adjacent to a granite source.

F. Photomicrograph of Feldspar in Basalt (length of field = 8 mm). Many fine-grained, dark-colored igneous rocks (basalts) contain large amounts of plagioclase in crystals too small to be seen in hand specimens. Under the microscope, however, the plagioclase feldspars appear as elongated rectangular crystals in a felt-like fabric.

FIGURE 2.6
The Feldspar Group

CALCITE AND DOLOMITE

Calcite is a major rock-forming mineral and appears in a great variety of forms. It occurs as the major mineral constituent in rock formations of limestone, chalk, and marble. It is also commonly deposited in caves, in hot and cold springs, and in veins. *Regardless of form, the following three physical properties distinguish calcite from other minerals: (1) perfect rhombohedral cleavage, (2) hardness of 3, and (3) reaction with dilute hydrochloric acid.* Calcite is thus one of the easiest minerals to identify. The major problem for beginning students is to become familiar with the variety of forms and colors in which calcite occurs.

Most calcite is white, although various impurities may tint it almost any color, or even black. Colorless, clear calcite shows strong double refraction.

A. Crystals of Calcite are extremely variable in form. They may be tabular, hexagonal, pyramidal, or rhombohedral. Large crystals may develop in caves, veins, and other voids, but calcite in most limestone usually occurs as compact aggregates of very small crystals. *(Jeffrey A. Scovil)*

B. Rhombohedrons of Clear Calcite. The perfect rhombohedral cleavage of calcite is one of its most distinctive properties. Regardless of crystal form or mode of occurrence, all calcite cleaves perfectly in three directions, not at right angles.

C. Crystalline Limestone. Calcite is the dominant mineral in limestone, and in some formations is the only mineral present. It may occur as interlocking crystals that range in size from submicroscopic to over an inch long. Additional forms of calcite in limestone are discussed and illustrated on pages 48–49.

D. Photomicrograph of Calcite (length of field = 8 mm). A mineral's physical properties are commonly well expressed on a microscopic scale. In this limestone specimen, perfect rhombohedral cleavage is clearly visible in the large interlocking grains.

E. Fractures Filled with Calcite. Calcite is one of the most soluble of the common minerals. It is dissolved readily by both surface and groundwater, and may precipitate subsequently from solution to fill cavities and fractures in rock formations. In this specimen, both the rock (limestone) and the fracture fill are composed of calcite.

F. Dolomite ($CaMg[CO_3]_2$) is a carbonate mineral similar in many respects to calcite. Crystals are commonly rhombohedral and occur in compact aggregates. Pure dolomite is usually pink but may be colorless, white, gray, green, or black. Dolomite is distinguished from calcite by its effervescence in HCl in powdered form only.

FIGURE 2.7
Calcite and Dolomite

19

OTHER ROCK-FORMING MINERALS

Specimens of other important rock-forming minerals are shown in Figure 2.8A–L. These include the ferromagnesian minerals, evaporite minerals, and clay minerals.

The ferromagnesian minerals (Figure 2.8 A–I) are complex silicates that contain appreciable amounts of iron and magnesium. They are generally dark green to black in color and have a high density. *The most important minerals in this group are olivine, pyroxenes, amphiboles, micas, chlorite, garnet, talc, and serpentine.*

Halite and gypsum (Figure 2.8J–K) are the two most common minerals formed by the evaporation of seawater or saline lake water and constitute important formations in some sequences of sedimentary rocks.

The clay minerals constitute the major part of shales and soils. They result from the alteration of aluminum silicates, especially feldspars. More than a dozen clay minerals can be distinguished; kaolinite (Figure 2.8L) is a common variety.

A. Olivine ranges in composition from Mg_2SiO_4 to Fe_2SiO_4. A complete solid solution of Mg and Fe exists in the crystal structure. The most common varieties of olivine are rich in Mg and are characterized by an olive-green color and by equidimensional glassy grains that lack cleavage. Olivine usually occurs in aggregates of small crystals.

B. Pyroxenes are complex silicates that contain substantial amounts of Ca, Mg, and Fe. They range in color from green to black or brown. Crystals are short and stubby, and when seen in cross section, they appear nearly square. Two cleavage planes intersect at right angles. Both the cleavage and crystal forms are generally poorly expressed in hand specimens but are well defined microscopically. Pyroxenes are essential constituents of dark-colored igneous rock rich in iron and are common in very high-grade metamorphic rock. *(Laurel Casjens, Utah Museum of Natural History, University of Utah)*

C. Amphiboles are complex silicates with a composition similar to that of pyroxene, but the amphibole crystal structure contains H_2O. Many physical properties of the two mineral groups are similar. Amphiboles are distinguished by long, columnar crystals and by visible cleavage in two directions intersecting at 56 and 124 degrees. Amphiboles (in particular the variety known as hornblende) are common as phenocrysts in many volcanic rocks. *(Laurel Casjens, Utah Museum of Natural History, University of Utah)*

D. Muscovite is a variety of mica, a distinctive mineral group with cleavage so perfect in one direction that most specimens can be separated into sheets thinner than paper. It is typically colorless and transparent in thin sheets.

E. Biotite. In addition to perfect cleavage, biotite mica is characterized by a pearly to vitreous luster. Cleavage sheets are elastic and flexible, and the mineral is notably soft. Micas are common in light-colored granitic rocks, and to a lesser extent in light-colored volcanics. They are also widespread in a variety of low- to intermediate-grade metamorphic rocks and in impure sandstones.

F. Chlorite is a common metamorphic mineral that results from the alteration of silicates such as pyroxene, amphibole, biotite, or garnet. It is characteristically dark green in color and occurs in aggregates of minute crystals or in finely disseminated particles. Individual crystals have perfect cleavage in one direction, which produces small, scaly, nonelastic plates.

FIGURE 2.8
Other Rock-Forming Minerals

G. Garnet. The garnet mineral group includes a series of silicate minerals that characteristically occur in well-formed, nearly spherical, 12-sided crystals. They have a hardness of 7–7.5 and are commonly red—although brown, yellow, green, and black varieties do occur, depending on composition. Garnets have no cleavage but may fracture along subparallel surfaces. The dominant fracture, however, is conchoidal. Most garnets are translucent to transparent and have a vitreous luster. They are common minerals in metamorphic rocks. *(Jeffrey A. Scovil)*

H. Talc is a secondary mineral formed from the alteration of magnesium silicates such as olivine, pyroxene, and amphiboles. It is distinguished from other minerals by its extreme softness (1 on the Mohs hardness scale), greasy feel, and perfect cleavage in one direction. Talc usually occurs as white to green foliated masses. Cleavage flakes are nonelastic.

I. Serpentine is usually compact and massive and often multicolored. A fibrous variety (asbestos) consists of delicate, fine, parallel fibers that can be separated easily. It is usually found in veins.

J. Halite is common salt (NaCl). It is characterized by cubic crystals that are usually clear and transparent, but impurities may impart reddish hues. It has perfect cleavage in three directions at right angles and a hardness between 2 and 2.5. Halite is by far the most common water-soluble mineral. It is precipitated from seawater or from saline lakes by evaporation and accumulates as salt beds interstratified with gypsum, silt, or shale. Salt layers may flow under pressure and may invade a weak area in an adjacent bedding plane to form plug-like bodies of solid salt.

K. Gypsum is a common sedimentary mineral formed by the evaporation of seawater or saline lakes. It is generally colorless to white and has a glassy to silky luster. Crystals are tabular. Gypsum is further characterized by its softness (2 on the Mohs scale), and by three unequal cleavage planes, one of which is perfect. Cleavage sheets are nonelastic. Varieties of gypsum include satin spar (fibrous gypsum with silky luster), alabaster (aggregates of fine crystals), and selenite (large, colorless, transparent crystals).

L. Kaolinite is one of several clay minerals that form from the decomposition of aluminum silicates, especially feldspar. It occurs in earthy aggregates of nearly submicroscopic, platy crystals. In hand specimens, it appears clay-like and is usually chalk white, although it may be stained red, brown, or black. It has a dull, earthy luster and disaggregates easily. Masses of kaolinite can be cut or shaped with a knife.

A. Pyrite is a yellow, metallic mineral composed of FeS_2. It commonly crystallizes in cubes, has a hardness of between 6 and 6.5, and breaks with a conchoidal fracture.

B. Chalcopyrite is a metallic mineral composed of copper, iron, and sulfur. It is characterized by a brassy yellow color, which often appears tarnished, and is slightly iridescent.

C. Galena (lead sulfide) is the most important lead mineral. It commonly crystallizes in cubes and is recognized easily by its perfect cubic cleavage and high density (7.5).

D. Graphite is composed of carbon atoms united in a sheet-like structure. It commonly occurs as foliated masses and is distinguished by extreme softness, perfect cleavage in one direction, and a greasy feel.

E. Sphalerite is the most important zinc mineral. It is difficult to recognize because of its extremely variable color (yellow, red, or brown to black). Its most diagnostic properties are a resinous luster and perfect cleavage in six directions. Note how the various cleavage planes reflect light in this specimen.

F. Native Copper usually crystallizes in poorly formed cubes, thus producing an irregular mass, twisted and wire-like in form. It has a copper color and metallic luster and is highly ductile and malleable. *(Jeffrey A. Scovil)*

G. Magnetite is a common iron ore and is the only mineral that is strongly magnetic. The magnetic property of this specimen is shown by the manner in which it attracts hair-like iron filings. Note how the alignment of the filings conforms to the specimen's north and south poles.

H. Hematite is an iron mineral that can vary greatly in appearance. Three varieties are common: (1) oolitic hematite, (2) specular hematite (metallic luster), and (3) earthy hematite. All are characterized by a deep red streak.

I. Limonite is a yellowish-brown, hydrous iron oxide that results from the alteration or weathering of iron minerals. It is amorphous and occurs commonly in both earthy and metallic varieties. The most diagnostic feature of limonite is its yellowish-brown streak.

FIGURE 2.9
Metallic Minerals

PROBLEMS

1. Examine each mineral specimen provided for this exercise by your instructor and briefly describe its physical properties as accurately as possible.

It is important to study a mineral specimen systematically and to record all the properties observed, even though they may seem obvious. An effective approach follows:

> **a.** Determine the type of luster, the color, and the streak.
>
> **b.** Determine the type of crystal or the texture of crystal aggregates.
>
> **c.** Test for hardness, cleavage or fracture, and other physical properties.

Do not break the specimens without your instructor's permission. After you have recorded all the physical properties observed, identify the mineral by referring to the mineral identification chart on pages 14–15. Note that Figure 2.4 is arranged systematically. The minerals are first grouped in three categories on the basis of luster and color. A second grouping is made on the basis of hardness, and a third on the basis of cleavage or fracture. Once you categorize a specific mineral according to these properties, you can then identify it from the brief description of that mineral's other characteristics.

For example, if you observe that a mineral is light colored, has a nonmetallic luster, is softer than steel or glass but will scratch a penny, and has good cleavage in four directions, you can identify it by referring to the chart's Nonmetallic Luster, Light Color section. Next, you would look under the list of minerals that are softer than glass and select the one that has four good cleavage directions. The only mineral possessing all of these properties is fluorite.

Photographs and descriptions of some of the common minerals are shown in Figures 2.5–2.9. When you have identified a mineral specimen, compare it with its picture in these figures. Note the similarities and determine which are significant physical properties. *The colored photographs are not a key to mineral identification and should not be used as such.* They are intended only as a study aid and reference.

2. Explain the law of constancy of interfacial angles.

3. List the important types of physical properties of minerals.

4. Give a complete and accurate definition of a mineral.

5. What is a solid solution?

6. List the rock-forming minerals in which solid solution is important.

7. What is the difference between a crystal face and a cleavage plane?

8. Give a brief definition of the term *hardness*.

9. What common mineral will react with dilute HCl?

10. What is the typical crystal form of quartz?

11. Are hand specimens of microcrystalline quartz single crystals or aggregates of crystals?

12. What is the most common occurrence of chert?

13. Is there a fundamental difference between milky quartz, smoky quartz, and rose quartz?

14. What are the diagnostic physical properties of the feldspars?

15. What are the two main subgroups of the feldspars? How do they differ?

16. How would you distinguish between Ca-plagioclase and K-feldspar?

17. What are the important physical properties of calcite?

18. How can microcrystalline calcite be recognized in a hand specimen?

19. What are the diagnostic properties of olivine?

20. In what rocks is olivine an important constituent?

21. How can you distinguish between amphibole and pyroxene?

22. What are the diagnostic properties of halite?

23. What are the diagnostic properties of gypsum?

24. Is a hand specimen of kaolinite an aggregate of crystals or a single crystal?

25. List the important rock-forming minerals.

26. Why would misidentification be likely if you attempted to identify unknown samples simply by comparing them with the photographs on pages 16–22, instead of testing their physical properties and using the mineral identification chart?

IGNEOUS ROCKS

■ OBJECTIVE

To recognize the major types of igneous rocks and to understand the genetic significance of their texture and composition.

■ MAIN CONCEPT

Igneous rocks can be classified on the basis of composition and texture. The composition of a rock provides information about the magma from which it formed, and about the tectonic setting in which it originated. The texture of a rock gives important insight into the cooling history of the magma.

■ SUPPORTING IDEAS

1. Igneous rocks are composed of silicate minerals, the most important of which are (a) plagioclase, (b) K-feldspar, (c) quartz, (d) mica, (e) amphibole, (f) pyroxene, and (g) olivine.

2. The major textures in igneous rocks are (a) phaneritic, (b) porphyritic-phaneritic, (c) aphanitic, (d) porphyritic-aphanitic, (e) glassy, and (f) pyroclastic.

3. Silicic magmas originate at subduction zones by partial melting of the oceanic crust.

4. Mafic magmas originate by partial melting of the upper mantle at divergent plate margins.

COMPOSITION OF IGNEOUS ROCKS

Approximately 99% of the total bulk of most igneous rocks is made up of only eight elements: oxygen, silicon, aluminum, iron, calcium, sodium, potassium, and magnesium. Most of these elements occur in the crystal structures of the rock-forming silicate minerals to produce feldspars, olivines, pyroxenes, amphiboles, quartz, and mica. These six minerals constitute over 95% of the volume of all common igneous rocks and are therefore of paramount importance in studying their classification and origin.

Magmas

Magmas rich in silica and aluminum are referred to as *silicic;* they tend to produce more quartz, potassium feldspar, and sodium plagioclase and generally form light-colored rocks. Magmas rich in iron, magnesium, and calcium are referred to as *mafic;* they produce greater quantities of olivine, pyroxene, amphibole, and calcium plagioclase. The resulting rocks are dark colored because of an abundance of the dark ferromagnesian minerals. (There are,

however, many exceptions to such generalizations, and the mineral composition of igneous rocks can only be roughly approximated by an observation of color.)

Crystallization

Crystallization of minerals from a magma occurs between 600 and 1200 °C. Those minerals with the highest freezing point crystallize first and thus develop well-formed crystal faces. Minerals that crystallize at lower temperatures are forced to grow in the spaces between the earlier-formed crystals and are commonly irregular in shape with few well-developed crystal faces. From laboratory studies of artificially produced magmas, and from petrographic studies of igneous rocks, a general order of crystallization has been established. This sequence is summarized in Figure 3.1 and is fundamental to the study of igneous rocks.

It is apparent from Figure 3.1 that in a mafic magma, olivine and Ca-plagioclase are the first minerals to form, followed by pyroxenes, amphiboles, and Na-plagioclase. Such a magma crystallizes between 900 and 1200 °C and produces rocks of the

gabbro-basalt family. In magmas rich in silica and aluminum, biotite and quartz form first, followed by K-feldspar and muscovite. Rocks of the granite-rhyolite family develop from such a magma at temperatures below 900 °C.

TEXTURE OF IGNEOUS ROCKS

Texture refers to the size, shape, and boundary relationships of adjacent minerals in a rock mass. In igneous rocks, texture develops primarily in response to the magma's composition and rate of cooling. Magmas located deep in Earth's crust cool slowly. Individual crystals grow to a more or less uniform size and may be more than an inch in diameter. In contrast, a lava extruded at Earth's surface cools rapidly, so the mineral crystals have only a short time in which to grow. The crystals from such a magma are typically so small that they cannot be seen without the aid of a microscope; the resultant rock appears massive and structureless. Regardless of crystal size, the texture of most igneous rocks is distinguished by a network of interlocking crystals.

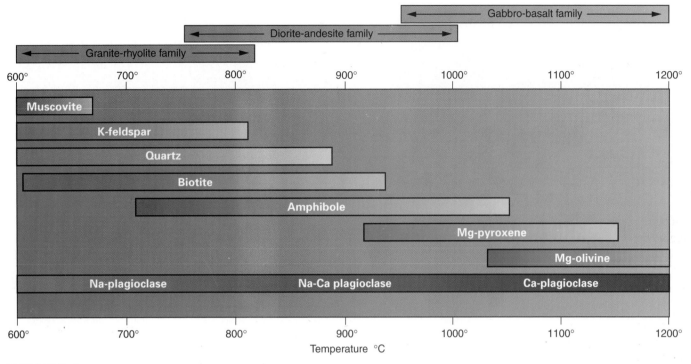

FIGURE 3.1
Order of Crystallization of Common Rock-Forming Minerals

Igneous rock textures are divided into the following types: (1) phaneritic, (2) porphyritic-phaneritic, (3) aphanitic, (4) porphyritic-aphanitic, (5) glassy, and (6) pyroclastic (Figures 3.2–3.7).

Phaneritic Texture

In phaneritic textures, individual crystals are large enough to be visible to the naked eye (Figure 3.2A and B). The grains in each specimen are approximately equal in size and form an interlocking mosaic. The size of crystals in a phaneritic texture can range from those barely visible, to crystals of more than an inch in length. Phaneritic texture develops from magmas that cool slowly and commonly develops in intrusive igneous bodies such as batholiths or stocks. Very coarse phaneritic rocks, in which the crystals are several feet long, almost invariably are found in large veins.

Porphyritic-Phaneritic Texture

A porphyritic-phaneritic texture is characterized by two distinct crystal sizes, both of which can be seen with the naked eye. The smaller crystals consti-

tute a matrix, or groundmass, that surrounds the larger crystals, or *phenocrysts* (Figure 3.3).

Aphanitic Texture

In aphanitic texture, individual crystals are so small they cannot be detected without a microscope (Figure 3.4). Rocks with this texture therefore appear to be massive and structureless. When a thin section of an aphanitic rock is viewed under a microscope, however, its crystalline structure is readily apparent. The rock is seen to be composed of numerous small crystals and, usually, some glass. Other examples of aphanitic texture viewed under a microscope are shown in the photomicrographs in Figures 3.10A and 3.11A.

Porphyritic-Aphanitic Texture

A porphyritic-aphanitic texture is defined as the texture of a rock with an aphanitic matrix in which embedded phenocrysts make up more than 10% of the total rock volume (Figure 3.5A and B). The phenocrysts are visible to the unaided eye. When phenocrysts are abundant, the rock at first glance may

appear to be phaneritic. Careful study of the area between the phenocrysts will indicate whether the matrix is aphanitic or phaneritic.

Glassy Texture

Glassy texture is similar to that of ordinary glass. It may occur in massive units (Figure 3.6A) or in a thread-like mesh resembling spun glass (Figure 3.6B). Crystals cannot be discerned in a glassy texture, even when the specimen is viewed under high magnification.

Pyroclastic Texture

Pyroclastic textures consist of broken, angular fragments of rock material (Figure 3.7A and B). In pyroclastic rocks, the fragmental material is composed of pumice, glass, droplets of lava, and broken crystals. Some sorting and stratification are generally present. Material finer than 4 mm is known as *tuff*. Material larger than 4 mm is referred to as *volcanic breccia*. If the fragments are exceptionally hot when deposited, they may fuse or weld together to form a dense mass.

A. Phaneritic Texture in this specimen is produced by relatively large grains of quartz (glassy grains) and feldspar (porcelainlike grains). The grains are about 1 cm across and are intergrown, so well-developed crystal faces have not formed. The black grains are biotite.

B. Fine-Grained Phaneritic Texture in this specimen is produced by small grains of quartz, white feldspar, and black mica. The grains of feldspar are less than 0.25 in. long.

FIGURE 3.2
Phaneritic Textures

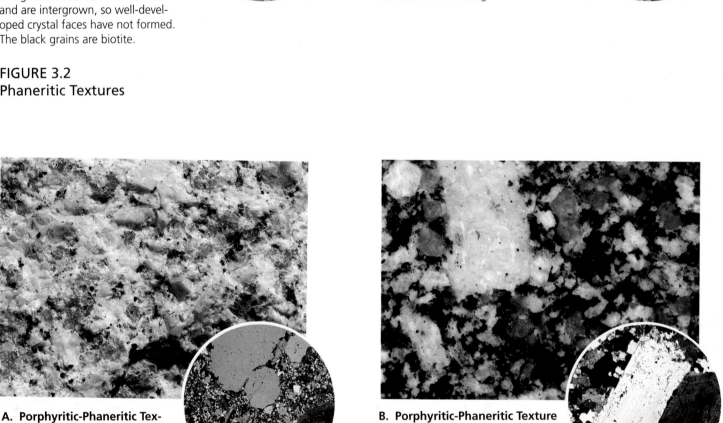

A. Porphyritic-Phaneritic Texture in this specimen consists of phenocrysts of large, rectangular crystals of pink K-feldspar. The matrix is composed of much smaller grains of white porcelain-like plagioclase, smoky quartz, and black ferromagnesian minerals.

B. Porphyritic-Phaneritic Texture in this specimen consists of phenocrysts of well-formed, pink potassium feldspar crystals set in a matrix of smaller white plagioclase and darker minerals.

FIGURE 3.3
Porphyritic-Phaneritic Textures

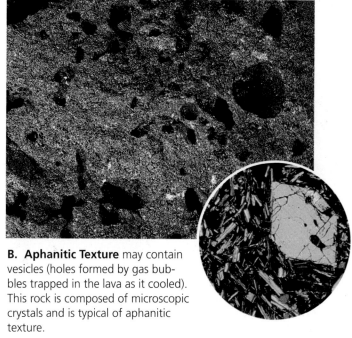

A. Aphanitic Texture may appear dense and structureless in hand specimens, such as this, but under the microscope individual mineral crystals can be distinguished (See also Figure 3.11A). Note that some secondary crystals have grown in cavities, so a few crystal specks can be seen without a microscope.

B. Aphanitic Texture may contain vesicles (holes formed by gas bubbles trapped in the lava as it cooled). This rock is composed of microscopic crystals and is typical of aphanitic texture.

FIGURE 3.4
Aphanitic Textures

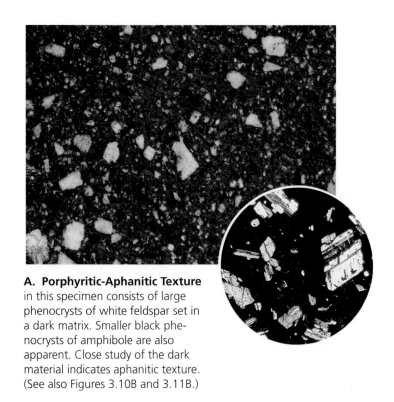

A. Porphyritic-Aphanitic Texture in this specimen consists of large phenocrysts of white feldspar set in a dark matrix. Smaller black phenocrysts of amphibole are also apparent. Close study of the dark material indicates aphanitic texture. (See also Figures 3.10B and 3.11B.)

B. Porphyritic-Aphanitic Texture in this specimen has small, white rectangular plagioclase crystals set in a dense red aphanitic matrix.

FIGURE 3.5
Porphyritic-Aphanitic Textures

A. Glassy Texture in igneous rocks is similar to common glass. It can be dense and massive, as in this specimen, or made of interwoven thread-like filaments, as shown in (B). Note the conchoidal fracture and knife-sharp edges.

B. Glassy Texture in this specimen occurs as tangled, thread-like filaments. Pore spaces are produced by gases that escaped as the lava cooled.

FIGURE 3.6
Glassy Textures

A. Pyroclastic Texture in this specimen consists of white, fine-grained volcanic ash with larger fragments of brown pumice and some angular rock fragments. An ash fall produced this texture.

B. Pyroclastic Texture produced by an ash flow is commonly dense and massive. Fragmental material in this specimen was very hot when extruded and flowed en masse close to the ground. As the material cooled, the hot particles fused. Black lenses were originally frothy pumice fragments like those in (A) but were so hot that the cellular structure collapsed and the mass became welded into a dense black lens.

FIGURE 3.7
Pyroclastic Textures

CLASSIFICATION AND IDENTIFICATION OF IGNEOUS ROCKS

Classification

The most useful and significant classification system for igneous rocks is based on two criteria: composition and texture. These criteria are important, not only in describing the rock so that it can be distinguished from other rock types, but also in drawing important implications about the rock's origin. A chart (Figure 3.8) in which variations in composition are shown horizontally and variations in texture are shown vertically provides an effective framework for classifying and naming igneous rocks.

Three major mineralogical criteria are used to classify igneous rocks:

1. *Presence or absence of quartz.* Quartz is an essential mineral in silicic rocks (rocks high in silica [Si]) and an accessory mineral in intermediate or mafic rocks—rocks high in magnesium (Mg) and iron (Fe). (In the classification of igneous rocks, an "essential mineral" is a mineral on which the rock classification is based. An accessory mineral occurs in minor amounts and is not a factor in classification.)

2. *Composition of the feldspars.* Potassium feldspars and sodium plagioclase are essential minerals in silicic rocks but are rare or absent in intermediate and mafic rocks. Calcium plagioclase is characteristic of mafic rocks.

3. *Proportion and kinds of ferromagnesian minerals.* As a general rule, mafic rocks are rich in ferromagnesian minerals, whereas silicic rocks are rich in quartz. Olivine is generally restricted to mafic rocks. Pyroxenes and amphiboles are present in mafic to intermediate rocks. Biotite is common in intermediate and silicic rocks.

The graph at the top of Figure 3.8 is not as complex as it may first appear. It shows the range in composition of the major igneous rock types. Each color zone represents a rock family (yellow—the granite-rhyolite family; yellow green—the diorite-andesite family; green—the gabbro-basalt family; and dark green—the peridotite family). The area on the graph allotted to each mineral represents the percentage of the rock composed of that mineral. For example, peridotite may be composed entirely of olivine, but pyroxene can occur in amounts up to 30% and Ca-plagioclase in amounts up to 15%. In the granite-rhyolite family, K-feldspar may form over 25% of the rock (the extreme left side of the graph). Note how the percentage of K-feldspar decreases as quartz increases; at the right margin of the granite-rhyolite zone, both quartz and K-feldspar decrease and plagioclase content increases.

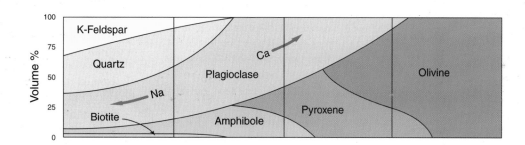

FIGURE 3.8
Classification of Igneous Rocks

From Figure 3.8 we can make the following summary of the composition of the igneous rock families:

The Granite-Rhyolite Family. The granite-rhyolite family (Figure 3.9) is characterized by the following mineral composition:

Quartz	10–40%
Potassium feldspar	15–30%
Plagioclase (high Na)	0–33%
Biotite and amphibole	8–15%

These rocks are commonly referred to as silicic. The magmas that produce them are high in potassium, silicon, and sodium, and are low in iron, magnesium, and calcium. Granites and rhyolites are therefore characteristically light colored.

The Diorite-Andesite Family. The diorite-andesite family (Figure 3.10) is intermediate in composition between the granite-rhyolite and gabbro-basalt families. It is characterized by the following composition:

Plagioclase	55–70%
Amphibole and biotite	15–40%

Plagioclase is approximately 50% albite and 50% anorthite. Potassium feldspar and quartz are present in minor amounts only. The diorite-andesite family is therefore characteristically gray in color.

The Gabbro-Basalt Family. The gabbro-basalt family (Figure 3.11) has the following composition:

Plagioclase (high Ca)	25–70%
Ferromagnesian minerals (olivine, pyroxene, and amphibole)	25–75%

These rocks crystallize from magmas that are relatively high in iron, magnesium, and calcium, but deficient in silica. Rock coloration is characteristically black or dark green.

The Peridotite Family. The peridotite family (Figure 3.12) is characterized by the following mineral composition:

Olivine	65–100%
Pyroxene	0–25%
Ca-plagioclase	0–5%
Ore minerals (i.e., magnetite, ilmenite, chromite)	0–10%

Hints on Mineral Identification in Hand Specimens

Identifying the major rock types in Earth's crust may be difficult for you at first because the most obvious properties we use to distinguish most things in the physical world (color, size, shape, and form) are not important in identifying rock type. Indeed, the specific color, size, and shape of a given rock have little significance. The rock properties that permit us to discriminate between the various igneous rock types are mineral composition and texture.

In a way, identifying rocks is like identifying woven fabric of cloth. The color and shape of a piece of fabric are insignificant. The composition of its threads—their size and the way they are woven—determines the nature of the fabric. It is much the same with rocks. A specimen's color, size, and shape are insignificant properties. Mineral composition and texture are the properties that distinguish one rock type from another, and provide information about their genesis. A student of geology, then, must learn to ignore a rock specimen's more obvious features—specific color, shape, and size—and to concentrate instead on composition and texture.

The following is a brief summary of the characteristics of the important minerals in igneous rocks.

Quartz. Occurs as irregular, *glassy* grains, commonly clear to smoky appearance; it has no cleavage.

Muscovite. Brass-colored flakes associated with quartz or K-feldspar. Perfect cleavage in one direction.

K-Feldspar. Porcelain luster. Commonly colored pink, white, or gray. Cleavage in two directions at right angles may be detected by a reflection of light when specimen is rotated.

Plagioclase. Usually gray or white in granite, dark-bluish color in gabbro. Striations common. Two cleavage directions at right angles may be detected.

Biotite. Small black flakes. Perfect cleavage in one direction. Reflects light.

Amphibole. Long, black crystals in a light-colored matrix. Cleavage at 60 and 120 degrees.

Pyroxene. Short, dull, greenish-black minerals in darker rocks. Cleavage in two directions at 90 degrees.

Olivine. Glassy, light-green grains.

PHOTOGRAPHS OF ROCK SPECIMENS

The photographs on the following pages illustrate the characteristics of the major igneous rock types as seen in hand specimens and under the microscope. The hand specimens are actual size. Some show a polished surface and others a natural fractured surface. All microscopic views show a magnification of approximately twenty times actual size. *Remember, the photographs are not a key to rock identification. Use them as a visual reference only for examples.* Note that the specimens in Figures 3.9–3.12 are arranged in families corresponding to the classification chart in Figure 3.8.

Feldspar
and quartz

A. Rhyolite is an aphanitic rock with the same composition as granite. It is commonly white, gray, or pink in color and almost always contains a few phenocrysts of feldspar or quartz (2–10%). If phenocrysts constitute more than 10% of the volume, the rock is properly termed porphyritic rhyolite. Because the texture is aphanitic, the only minerals that can be identified in a rhyolite hand specimen are those occurring as phenocrysts. This specimen is from a typical porphyritic rhyolite deposit in the western United States. The dark phenocrysts are smoky quartz, but most of the rock has a characteristic aphanitic texture. In thin section, flow structures are commonly apparent, and a considerable amount of glass is present between the fine-grained crystals. The specimen in Figure 3.4A is also a rhyolite.

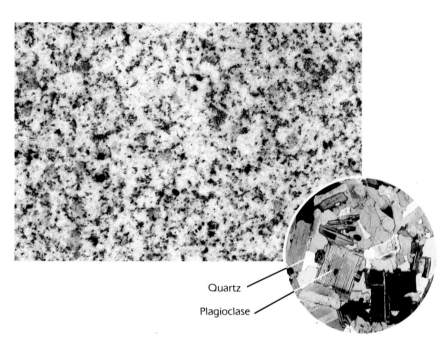

Quartz

Plagioclase

B. Granite is a phaneritic igneous rock composed predominantly of feldspar and quartz. Biotite, amphibole, and plagioclase (early-formed crystals) are generally euhedral (having well-developed crystal faces). In this specimen, white crystals are sodium-rich plagioclase, light-gray glassy grains are smoky quartz, and dark crystals are biotite. Many granites are gray, but if K-feldspar dominates, coloration may be pink or red. Specimens in Figures 3.2A and B are also granites. From one granitic body to another, there may be a wide range in the average crystal size, but the crystals are large enough to be seen without the aid of a microscope.

K-feldspar

C. Porphyritic Granite usually indicates two stages of cooling: an initial stage in which large crystals formed, followed by a stage in which more rapid cooling developed smaller grains. In this specimen, the phenocrysts are large crystals of pink K-feldspar. The porcelain-white mineral is plagioclase, and the glassy grains are quartz. Small biotite grains are also apparent. Although grain size, color, and general specimen appearance are quite different from the specimen in (B), the composition of the two rocks is essentially the same. Both are composed mostly of quartz, K-feldspar, and calcium plagioclase. The only significant difference is in texture. The specimen in Figure 3.3A is also porphyritic granite.

FIGURE 3.9
Granite-Rhyolite Family

A. Andesite is an aphanitic rock composed of Na-plagioclase, amphibole, and pyroxene, with little or no quartz. Specimens are generally dark gray, green, brown, or red. The name *andesite* comes from the Andes Mountains in South America, where volcanic eruptions have produced this rock type in great abundance. Andesitic magmas originate in subduction zones by partial melting of the oceanic crust. After basalt, andesite is the most abundant volcanic rock. Completely aphanitic andesite is relatively rare because most flows contain some phenocrysts.

Feldspar

B. Porphyritic Andesite is the most common variety of intermediate extrusive rock. Phenocrysts are composed mainly of plagioclase, amphibole, or biotite set in an aphanitic matrix of plagioclase and some glass. In this specimen, black phenocrysts are amphibole. Porphyritic andesite and basalt comprise 95% of all volcanic material. The specimen in Figure 3.5B is also porphyritic andesite.

Pyroxene

Plagioclase

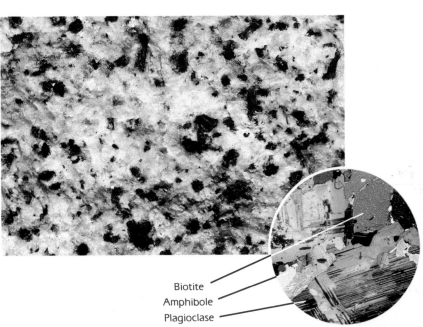

C. Diorite has a texture essentially the same as granite. (Compare this specimen with that in Figure 3.2.) The two differ in composition only. Essential minerals in a granite are K-feldspar, quartz, and Na-plagioclase, whereas diorite is composed predominantly of Na-Ca-plagioclase and ferromagnesian minerals. In this specimen, ferromagnesian minerals (amphibole) are abundant and give it a speckled appearance. In thin section, it is clear that quartz and K-feldspar are absent; plagioclase and amphibole predominate. Quartz accounts for less than 5% of the total volume. Diorite occurs in large intrusives such as stocks or batholiths. It is found also in dikes, sills, and laccoliths.

Biotite
Amphibole
Plagioclase

FIGURE 3.10
Diorite-Andesite Family

Plagioclase

A. Basalt is an aphanitic rock composed predominantly of calcium-rich plagioclase and pyroxene, with small amounts of olivine and amphibole. It is characteristically black, dense, and massive. Although individual crystals cannot be seen without magnification, basalt is generally easy to recognize. Many specimens are clinker-like in appearance, with half of the total volume consisting of small holes, or vesicles. Vesicular texture develops as gas bubbles rise toward the top of a flow and are trapped by the overlying cooling crust. Under the microscope, crystals of individual minerals can be seen and studied. Plagioclase crystals commonly occur as a mesh of lath-like crystals surrounding crystals of pyroxene, olivine, or amphibole.

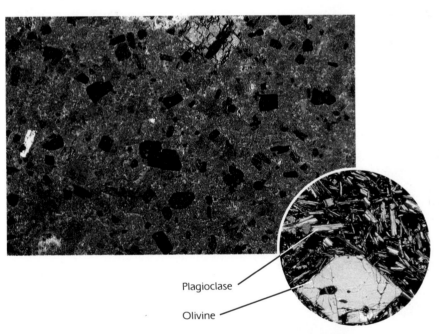

Plagioclase

Olivine

B. Porphyritic Basalt is common because many basaltic lavas have some early-formed crystals of olivine, pyroxene, amphibole, or Ca-plagioclase. In this specimen, phenocrysts are amphiboles. Figure 3.5A is also a porphyritic basalt, but phenocrysts are plagioclase. Basalt is the most abundant extrusive rock. It is the bedrock for the oceanic crust and is found also in great floods produced at rift systems on some continents. Basaltic magma forms from partial melting of the upper mantle and is extruded mostly along fissures in rift zones.

Olivine

Plagioclase

C. Gabbro has a phaneritic texture similar to granite but is composed almost entirely of pyroxene and Ca-plagioclase, with minor amounts of olivine. Its composition is thus the same as basalt. Calcium-rich plagioclase is the dominant feldspar and usually occurs as dark, elongated crystals, like those in this specimen. Coloration is characteristically dark green, dark gray, or almost black because of the predominance of Ca-plagioclase and ferromagnesian minerals. Gabbro is not common in the continental crust, but large exposures occur north of Duluth, Minnesota, in Labrador, and in Finland. Most gabbro is produced at divergent plate margins and forms the bottom layer of the oceanic crust.

FIGURE 3.11
Gabbro-Basalt Family

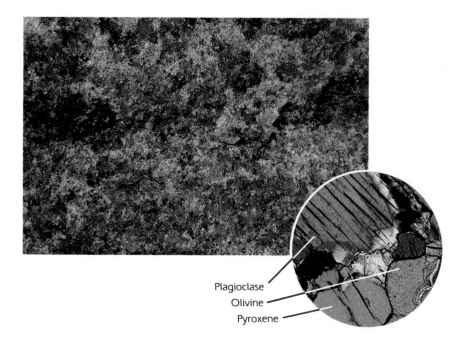

Plagioclase
Olivine
Pyroxene

A. Peridotite is relatively easy to identify because it is composed almost entirely of olivine; however, significant amounts of pyroxene and lesser amounts of Ca-plagioclase may occur. This specimen is composed mostly of olivine. Olivine crystals are characteristically light green and have a distinctive glassy luster. They are typically equidimensional and are medium to coarse grained, so the rock may resemble a green, glassy sandstone. Rocks of the peridotite family have phaneritic texture and originate far below the surface. No extrusive rocks have an equivalent composition.

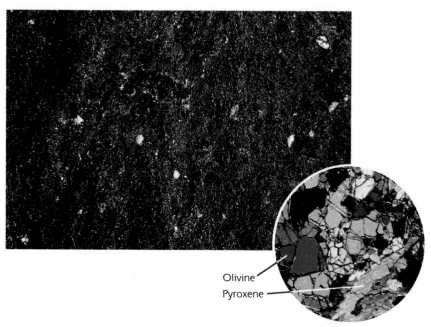

Olivine
Pyroxene

B. Peridotite (Kimberlite) may have a relatively fine-grained texture and may lack the glassy, green appearance of the specimen in (A), especially if pyroxene, Ca-plagioclase, and some accessory minerals are present. This specimen is a kimberlite, a variety of peridotite that is host rock for diamonds.

C. Peridotite is not an abundant rock type in Earth's crust, but it is important in studying Earth's dynamics because it is believed that the entire upper mantle is composed almost exclusively of this type of rock. Indeed, it is probable that the melting and movement of peridotite in the asthenosphere is intimately involved in the tectonic system.

FIGURE 3.12
Ultramafic Rocks

Flow lines

A. Obsidian is a massive volcanic glass. As shown in this specimen, it typically breaks with a conchoidal fracture and has a bright, glassy luster. Despite its composition (mostly SiO_2), it is characteristically jet black due to countless dust-like particles of magnetite or ferromagnesian minerals. More rarely, yellow, red, or brown hues are produced by oxidized magnetite or hematite. In thin section, flow lines are common, and the rock appears totally black under polarized light. Obsidian is not crystalline but does contain skeletal crystal embryos called crystallites. Many aphanitic rocks contain appreciable quantities of glass, which fills the space between crystals. The specimen in Figure 3.6A is also obsidian.

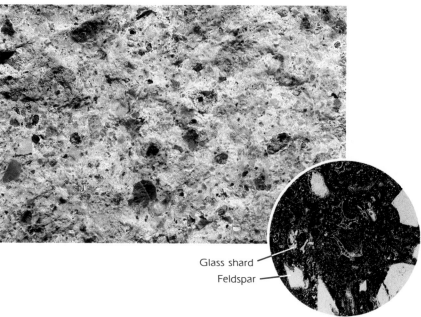

Glass shard
Feldspar

B. Tuff is composed of fragments of volcanic material and has a fragmental texture more characteristic of sedimentary rock. It forms from fragmental material and droplets of lava that erupt from volcanic vents and are transported through the air before settling back to the surface. This hot, fragmental material is referred to as pyroclastic. Rocks formed from the consolidation of such material are classified by grain size as follows: (1) *tuff* (fine ash and dust less than .25 inch in diameter), (2) *volcanic breccia* (coarse ash and angular blocks .25–2 inches in diameter), and (3) *agglomerates* (volcanic bombs and blocks greater than 2 inches in diameter). This specimen is composed of coarse fragments of pumice and fine-grained ash.

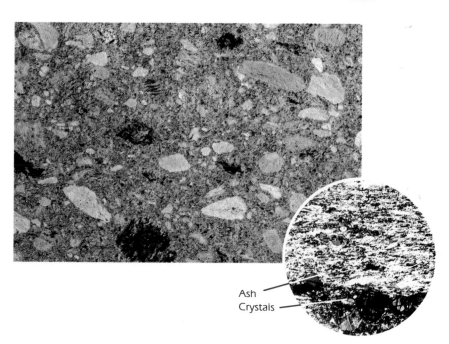

Ash
Crystals

C. Ash-Flow Tuff is composed of volcanic ash (fragments of volcanic glass, broken crystals, droplets of lava, and fragments of country rock fused into a tight, coherent mass). It differs from ash-fall tuff (B) in that the fragments were very hot when extruded. As a result, the fragments were fused, flattened, or bent out of shape. This unique texture forms from a cloud of hot, incandescent ash that moves very rapidly, like a flow. The specimen in Figure 3.7B is also an ash-flow tuff.

FIGURE 3.13
Other Igneous Rocks

Origin of Igneous Rocks

The origin of magma is not known from direct observations, but during the last few decades, our understanding of the chemistry and physics of liquid rock material has increased greatly. This knowledge is based on observations of synthetic magmas made in the laboratory, on the observation of volcanic products, and on studies of Earth's geophysical properties.

We know from seismic evidence that Earth is essentially solid to a depth of 2900 km, and that only its outer core is liquid. The liquid core, however, cannot be the source of magma. The core's density ranges from 9 to 10 gm/cm³, a density much greater than that of any magma or igneous rock in the crust. (Basalt, one of the densest igneous rocks, has a density of only 3 gm/cm³.) Magma must therefore originate from the localized melting of solid rock in the upper mantle and lower crust, at depths ranging from only 50 to 200 km below the surface.

To understand the origin of magma, remember that a rock does *not* have a specific melting point. Each mineral within the rock begins to melt at a different temperature. As shown in the diagram in Figure 3.1, muscovite, K-feldspar, and quartz not only crystallize in sequence at temperatures ranging from a little over 800 to 600 °C, but they begin to melt in this temperature range as well. Amphibole melts at a higher temperature, followed by the melting of pyroxene and olivine. Sodium-rich plagioclase begins to melt at about 600 °C, whereas calcium-rich plagioclase may not melt until it reaches a temperature of 1200 °C. In reality, melting and crystallization are complex physical and chemical processes. It is, however, well established that some minerals melt at lower temperatures than others. This fact is important in understanding the general way in which a magma originates.

In addition to temperature, remember that other important factors—such as pressure, amount of water present, and overall composition of the rock—influence melting. For example, melting can be initiated by either an increase in temperature or a decrease in pressure, and is enhanced by the amount of water present in the pore spaces of the rock.

The diagrams in Figure 3.14 summarize our present understanding of how and where a magma originates. Global studies of the distribution of flood basalts, volcanoes, batholiths, and associated mountain belts indicate that basaltic magma is generated at divergent plate margins and that granitic-andesitic magma is generated in subduction zones. In both areas, variations in temperature and pressure occur as a result of plate movement, and magma is produced by the partial melting of the lower crust or upper mantle.

Generation of Basaltic Magma

Basaltic magma is believed to be generated at divergent plate margins by partial melting of the asthenosphere (Figure 3.14A). The asthenosphere, as well as the lower mantle, is almost certainly composed of peridotite, a rock made up largely of the minerals olivine, pyroxene, and minor amounts of plagioclase. The balance between temperature and pressure in the asthenosphere is just about right for peridotite to begin to melt. Below the asthenosphere, the pressure is too great for melting to occur, and at more shallow depths above the asthenosphere, the temperature is too low.

The general process of how basaltic magma originates follows: As material in the asthenosphere moves slowly upward and outward, the lithosphere splits and moves apart. This reduces pressure above a section of the asthenosphere, and melting begins. Plagioclase melts first, followed by pyroxene, and ultimately by the melting of olivine. If only part of the mantle rock melts, much of the olivine remains solid, so the resulting magma is richer in those elements that compose plagioclase and pyroxene. In fact, that is the composition of basaltic magma. Scientists therefore hypothesize that basaltic magma originates from partial melting of the asthenosphere at rift zones, where pressure is reduced as a result of the splitting and spreading apart of lithospheric plates. A decrease in pressure thus plays an important role in the generation of basaltic magma. The basaltic magma, being less dense than the surrounding peridotite, rises along the spreading center and is extruded as basalt flows in the rift zone.

Generation of Granitic Magma

In a subduction zone, the basaltic oceanic crust and water-saturated oceanic sediments descend into the mantle. This mixture of basalt and sediment is heated by the friction between colliding lithospheric plates and by the higher temperatures at depth as the mixture is submerged (Figure 3.14B). As the basalt and oceanic sediments are heated, the silica-rich minerals begin to melt first, some at temperatures as low as 500 °C. Partial melting of the basaltic crust thus produces a magma richer in silica than the magma produced at divergent plate boundaries. (See Figure 3.1.) This silica-rich magma rises upward in the orogenic belt to produce granitic intrusions and andesitic-rhyolitic flows. The extreme pressure in the roots of a deformed mountain belt, at converging plate margins, would also increase the temperature enough to begin the melting of some of the minerals in metamorphic rocks. The liquid would rise, collect in larger bodies, and produce chambers of granitic magma.

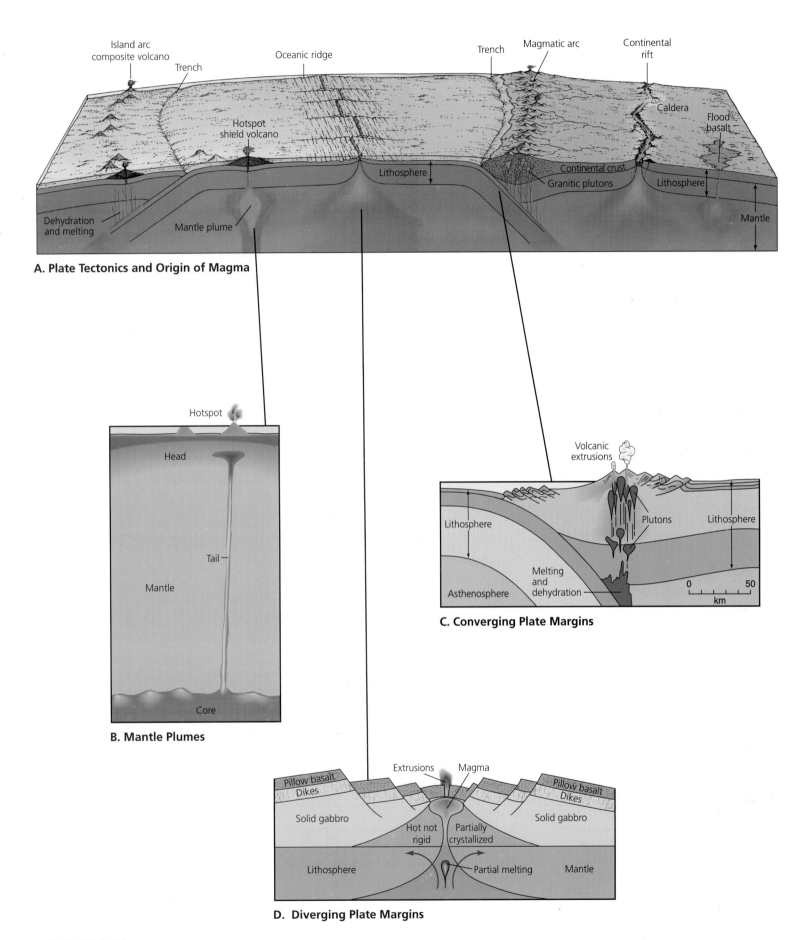

A. Plate Tectonics and Origin of Magma

Island arc composite volcano
Trench
Oceanic ridge
Trench
Magmatic arc
Continental rift
Caldera
Flood basalt
Hotspot shield volcano
Lithosphere
Continental crust
Granitic plutons
Lithosphere
Mantle
Dehydration and melting
Mantle plume

B. Mantle Plumes

Hotspot
Head
Tail
Mantle
Core

C. Converging Plate Margins

Volcanic extrusions
Lithosphere
Plutons
Lithosphere
Asthenosphere
Melting and dehydration
0   50
km

D. Diverging Plate Margins

Extrusions
Magma
Pillow basalt
Dikes
Pillow basalt
Dikes
Solid gabbro
Solid gabbro
Hot not rigid
Partially crystallized
Lithosphere
Partial melting
Mantle

FIGURE 3.14
The Origin of Magma

1. Identify specimens provided by your instructor. To do this efficiently, we suggest following these procedures:

a. Examine the rock and determine the type of texture. Then refer to the rock as (a) phaneritic, (b) porphyritic-phaneritic, (c) aphanitic, (d) porphyritic-aphanitic, (e) glassy, or (f) pyroclastic.

b. Determine the percentage of dark minerals in the rock. Then refer to the rock as (a) silicic—a few dark minerals, generally a light gray color; (b) intermediate—nearly 50% dark minerals, dark gray in color; or (c) mafic—over 70% dark minerals, very dark to black.

c. Determine the approximate percentage and type of feldspar: (a) pink feldspar is almost invariably a potassium feldspar; (b) white or gray feldspar may be either potassium feldspar or plagioclase. If the feldspar has striations, it is definitely plagioclase.

d. Determine the approximate percentage of quartz: (a) 20–40% quartz—granite-rhyolite family; (b) less than 20% quartz—diorite-andesite family; (c) no quartz—gabbro-basalt family.

e. Use the rock chart (Figure 3.8) and determine the rock name.

Example

(1) A phaneritic rock composed of 25% olivine, 50% pyroxene, and 25% plagioclase would be a gabbro. (See graph of mineral composition at top of Figure 3.8.)

(2) An aphanitic rock containing phenocrysts of amphibole in a light-colored matrix, presumably plagioclase, would probably be a porphyritic andesite. An aphanitic rock containing phenocrysts of quartz would be a porphyritic rhyolite.

(3) A phaneritic rock containing 15% amphibole, no quartz, and 85% plagioclase would be diorite.

Aphanitic rocks have few distinguishable crystals and are difficult to identify, with certainty, from a hand specimen alone. The following general guidelines will be useful in identifying aphanitic igneous rocks:

- If quartz phenocrysts are present, the rock is a rhyolite.

- If potassium feldspar phenocrysts are present, the rock is a rhyolite.

- If amphibole phenocrysts are present, the rock is an andesite.

- If pyroxene or olivine phenocrysts are present, the rock is a basalt.

- If phenocrysts are conspicuous, the texture is porphyritic, and the adjective *porphyritic* is added to the rock name (porphyritic granite or porphyritic basalt).

Note that some types of igneous rock do not fit well in the classification system used in Figure 3.8. The most important exceptions are described here.

Obsidian. A massive volcanic glass, usually jet-black due to the presence of dust-like particles of magnetite and ferromagnesian minerals. Most obsidian is rich in silica and has a chemical composition similar to that of granite and rhyolite.

Pumice. A porous volcanic glass with a texture consisting of subparallel glass fibers tangled together.

Tuff and Volcanic Breccia. A volcanic rock of pyroclastic texture, formed from consolidated volcanic ash. Tuff is fine grained, similar to fine sand or mud. Breccia is coarse grained, like a gravel. The composition of pyroclastic rocks is variable.

Ash-Flow Tuff. If the grains of ash, pumice, and crystal fragments in a pyroclastic rock are fused together, the term *ash-flow tuff* is used.

2. Compare your specimens with the illustrations and descriptions on pages 31–35. What are the distinguishing characteristics of each rock type?

3. How does your specimen differ from the same rock type illustrated in this manual? Is this difference a fundamental difference in texture and composition or a minor difference in a less diagnostic feature such as color or specimen shape?

4. Examine the photographs of the hand specimens and photomicrographs on the next two pages. Using the information on composition, determine the name of each rock and describe its origin.

5. In Hawaii, many basaltic flows are only 1–2 ft thick. On the Columbia Plateau, similar flows are 20–40 ft thick. What factor is important in determining the thickness of basalt flows?

6. Could phenocrysts be found in an obsidian? Explain your answer briefly.

7. What is the difference between granite and diorite?

8. What is the origin of vesicles in a basalt?

9. What is the difference between basalt and andesite?

10. Which rock type would be most likely to form a sill? A laccolith? Why?

11. What minerals are likely to form phenocrysts in a basalt? Why?

12. A granite may be pink, red, or various shades of gray, and may be coarse, medium, or fine grained. What characteristics are typical of all granites and distinguish granites from other rock types?

13. What rock type is most abundant in the oceanic crust?

14. Why are rocks classified on the basis of texture and composition instead of size, shape, and color?

15. What igneous rock types form in a mountain belt? Why are these rocks different from those formed at the oceanic ridge?

Specimen A

Composition
K-feldspar	40%
Quartz	30%
Biotite	5%
Amphibole	5%
Na-plagioclase	20%

Rock name _____

Origin _____

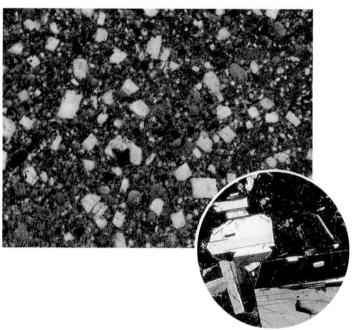

Specimen B

Composition
Ca-plagioclase	50%
Pyroxene	35%
Amphibole	15%

Rock name _____

Origin _____

Specimen C

Composition
Ca-Na-plagioclase	70%
Amphibole	25%
Biotite	5%

Rock name _____

Origin _____

Specimen D

Composition
K-feldspar	40%
Na-plagioclase	30%
Quartz	20%
Biotite	10%

Rock name _____

Origin_____

Specimen E

Composition
K-feldspar	30%
Quartz	35%
Biotite	5%
Amphibole	5%
Na-plagioclase	25%

Rock name _____

Origin_____

Specimen F

Composition
Ca-plagioclase	50%
Pyroxene	45%
Olivine	5%

Rock name _____

Origin_____

EXERCISE 4 SEDIMENTARY ROCKS

OBJECTIVE

To recognize the major types of sedimentary rocks and to understand the genetic significance of their texture and composition.

MAIN CONCEPT

Sedimentary rocks are classified on the basis of texture and composition. Two main groups are recognized: (1) clastic (formed from fragments of other rocks), and (2) nonclastic (formed by chemical or organic processes).

SUPPORTING IDEAS

1. Sedimentary rocks are composed of material that has been weathered, transported, and deposited by various processes operating at Earth's surface (running water, wind, and glaciation, for example).
2. Most minerals that form sedimentary rocks are relatively stable at surface temperatures and pressures.
3. The most important constituents in sedimentary rocks are (a) quartz, (b) calcite, (c) clay minerals, and (d) rock fragments.
4. The most important environments of sedimentation are (a) deltas, (b) shallow-marine environments, (c) beaches and bars, (d) deep-marine environments, and (e) reefs. Other important settings in which sedimentation occurs are (a) fluvial environments, (b) alluvial fans, (c) deserts, and (d) glaciers.

COMPOSITION OF SEDIMENTARY ROCKS

Because sediment can be derived from any preexisting source rock, we might expect the composition of sedimentary rocks to be extremely variable and complex. This is indeed true if the sediment is deposited close to the source area. If weathering and erosion are prolonged, however, sedimentary processes concentrate materials that are similar in size, shape, and composition into separate, distinct deposits. Most sedimentary rocks are thus composed of the materials that are abundant in other rocks and that are stable under the conditions of surface temperature and pressure. The great bulk of most sedimentary rocks is composed of only four constituents: (1) quartz, (2) calcite, (3) clay minerals, and (4) rock fragments.

Quartz

Quartz is one of the most abundant clastic minerals in sedimentary rocks. This is because quartz is one of the most abundant minerals in the granitic continental crust and because it is extremely hard, resistant, and chemically stable. Weathering processes decompose and disintegrate less stable minerals such as feldspars, olivine, pyroxene, amphibole, and mica. Quartz, however, remains unaltered and is transported by running water and is commonly concentrated as deposits of sand in fluvial and beach environments. Silica in solution, or in particles of colloidal size, is also a product of the weathering of igneous rocks and is commonly precipitated as a cement in certain coarse-grained sediments.

Calcite

Calcite is the major constituent of limestone and is the most common cementing material for sands and shales. Calcium is derived from igneous rocks that are rich in calcium-bearing minerals such as calcium plagioclase. The carbonate is derived from water and carbon dioxide. Calcium is precipitated as $CaCO_3$ or is extracted from seawater by organisms and concentrated as shell material. When the organisms die, the shell fragments commonly accumulate as clastic particles that ultimately form a variety of limestones.

Clay

Clay minerals develop from the weathering of silicates, particularly the feldspars. They are very fine grained and are concentrated in mud and shale. The abundance of feldspar in Earth's crust and the fact that it decomposes readily under atmospheric conditions accounts for the large amount of clay minerals in sedimentary rocks.

Rock Fragments

Fragments of the parent rock, in which constituent minerals have not disaggregated, are the major constituents of coarse-grained, clastic deposits such as gravel. Rock fragments are also present as grains in some sandstones; the parent rock is usually a basalt, slate, or some other fine-grained rock.

Other Minerals in Sedimentary Rocks

Deposits of quartz, calcite, and clay, either alone or in various combinations, form most sedimentary rocks. Other minerals are, however, sometimes abundant enough to form distinct strata.

Dolomite, $CaMg(CO_3)_2$, may replace calcite in limestone. Feldspars and mica may be concentrated in some sandstones if weathering, erosion, and deposition occur rapidly. Halite and gypsum are precipitated by the evaporation of salt lakes or seawater and in certain environments may accumulate in thick layers. Decayed plant material may accumulate in swamps and may become thick beds of coal. Corals and other types of marine organisms commonly form organic reefs that may develop a distinctive type of limestone.

TEXTURE OF SEDIMENTARY ROCKS

Texture in sedimentary rocks is significant because it provides important clues concerning the distance that the sediment has been transported and the environment in which it was deposited. Two basic texture types are recognized: (1) clastic (mineral fragments and rock debris) and (2) nonclastic (mostly crystals that have grown from solutions or, organic material).

Clastic Texture

Examples of clastic texture are shown in Figure 4.1. The size of the clastic particles in sedimentary rock ranges from large blocks that are many feet in diameter to fine dust. Particles are referred to as coarse grained (over 2 mm in diameter), medium grained (1/16–2 mm in diameter), or fine grained (less then 1/16 mm in diameter).

Rounding. Rock fragments and mineral particles may be rounded (Figure 4.1A) or angular (Figure 4.1B), depending on the amount of abrasion they have undergone. Sediment moved by ice or by the direct action of gravity is commonly angular, whereas particles carried by wind or water are rounded by continual abrasion. In general, grain size and shape are a rough measure of the distance of transport. For example, large, angular boulders indicate a nearby source area in a mountainous terrain because significant transport by streams rapidly rounds off the rock corners and wears down the boulder to gravel and sand size.

Sorting. Sorting refers to the separation of particles by size. It is a very

important textural characteristic because it can provide clues concerning the history of transportation and the environment in which the sediment accumulated. Layers of well-sorted material (Figure 4.1C–D) are composed of grains of one dominant size and composition. Such layering results only after considerable transport, during which particles similar in size and density are concentrated by prolonged current action. Poorly sorted material (Figure 4.1A) contains grains of several different sizes. Glaciers do not sort material but deposit coarse and fine particles together. Mud flows also produce poor sorting. Wind and water are the best sorting agents, concentrating materials of one dominant grain size in dunes, beaches, bars, and mud flats.

Compaction and Cementation.

After a sediment is deposited, it may be compacted and cemented to form solid rock. The dominant cementing materials are calcite, quartz, iron, and chert. These may be carried by groundwater from a foreign source or derived by solution activity from the site of sedimentation. The extent of cementation is an important textural characteristic in clastic rocks.

Nonclastic Texture

Examples of nonclastic texture are shown in Figures 4.2A–D.

Crystalline Texture. Crystalline texture is in marked contrast to the clastic texture just described. The minerals precipitated from seawater, groundwater, or lakes form this texture, which consists of a network of interlocking crystals similar to the texture in some igneous rocks. Crystalline textures are described as coarse (greater than 2 mm), medium (1/16–2 mm, Figure 4.2A), and fine or microcrystalline (less than 1/16 mm, Figure 4.2B). Deposits from springs and deposits formed in caves are commonly microcrystalline with a banded appearance that results from chemical variations and impurities during deposition.

Skeletal Texture. The skeletal texture shown in Figure 4.2C was formed by the accumulation of skeletal parts of invertebrate marine life. Calcium carbonate is removed from seawater by marine organisms to make their shells

and other hard parts. When the organisms die, their shell material settles to the sea floor and may be concentrated as shell fragments on a beach or near a reef. The texture of the resulting rock is similar to a clastic texture, but the material is unique in that it consists of the skeletal fragments of organisms that originate in the environment of deposition. Skeletal texture predominates in many limestones.

Oolitic Texture. Oolitic texture resembles sand, but the individual grains have concentric layers (Figure 4.2D). This layering is produced when calcium carbonate, precipitated on the sea floor, is agitated by wave or current action and accumulates around a tiny shell fragment or grain of silt. As the particle moves to and fro, thin, spherical layers are built up by accretion. Close examination of an oolitic texture reveals a concentric structure around each nucleus and, generally, minor amounts of associated shell debris. (See Figure 4.5C.)

PHOTOGRAPHS OF ROCK SPECIMENS

The photographs that follow illustrate the characteristics of the major types of sedimentary rocks, as seen in hand specimens and under a microscope. The hand specimens are actual size. Some have a polished surface, which shows textural characteristics better. Others have a natural fractured surface. All microscopic views show a magnification of approximately 20 times the actual size. *Remember, the photographs are not a key to rock identification!* Use them only as a visual reference for selected examples.

A. Coarse-Grained Clastic Texture consists of particles larger than 2 mm in diameter. In this specimen, the particles are well rounded but poorly sorted.

B. Coarse-Grained Angular Pebbles are important textural characteristics in this specimen. The angular particles indicate that little abrasion occurred and suggest a short distance of transport.

C. Medium-Grained Texture consists of particles from 1/16 to 2 mm in diameter—the size of common sand. In this specimen, the particles are well rounded and well sorted. (See also Figure 4.4E.)

D. Fine-Grained Texture consists of particles less than 1/16 mm in diameter, the grain size of mud. The individual particles are too small to be seen without high magnification. Many fine-grained clastic rocks are stratified with fine laminations, like the specimen shown here. (See also Figure 4.4F.)

FIGURE 4.1
Clastic Textures

A. Crystalline Texture in sedimentary rocks consists of a network of interlocking crystals and is similar in many respects to igneous rock textures. In this specimen, the crystals are all calcite, so the boundaries between individual crystals are not clearly defined. Under the microscope, however, crystal boundaries and cleavage are obvious. (See also Figure 4.5A.)

B. Microcrystalline Texture is similar to the texture of aphanitic igneous rocks. The grains are so small they can only be seen with high magnification. These tiny microscopic crystals of calcite make up the entire rock. (See also Figure 4.5B.)

C. Skeletal Texture consists of fragments of shells. Many different varieties of this texture occur, depending on the type and size of shell debris and the nature of the cementing material. Note that the sand-sized particles are not quartz but shell fragments.

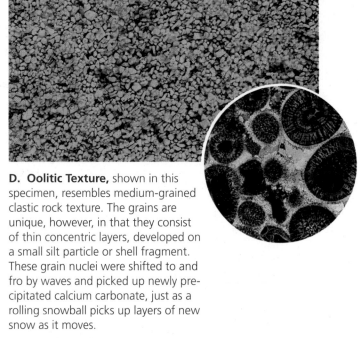

D. Oolitic Texture, shown in this specimen, resembles medium-grained clastic rock texture. The grains are unique, however, in that they consist of thin concentric layers, developed on a small silt particle or shell fragment. These grain nuclei were shifted to and fro by waves and picked up newly precipitated calcium carbonate, just as a rolling snowball picks up layers of new snow as it moves.

FIGURE 4.2
Nonclastic Textures

CLASSIFICATION OF SEDIMENTARY ROCKS

Although sedimentary processes produce a wide variety of rocks that differ in general appearance, most sedimentary rocks fit into three major categories: (1) clastic, (2) chemical, or (3) organic (Figure 4.3).

Clastic Rocks

Clastic rocks consist predominantly of fragments and debris of other rocks. These materials are as familiar as gravel, sand, and mud. Clastic rocks are thus classified according to grain size, with subdivisions based on composition (Figure 4.4A–F).

Chemical Rocks

Chemical rocks are precipitated directly from water, usually as a result of evaporation or changes in the chemistry of the water. Limestone and dolomite are the most abundant types of chemically formed rocks. Limestone contains more than 50% calcite. Other materials present in limestone may include clay, quartz, rock fragments, or iron oxide. The calcite that is precipitated chemically may form crystalline limestone, microcrystalline limestone, or oolitic limestone (Figure 4.5A–C). Gypsum and halite also form important evaporite deposits, and chert is common as a component in many limestone layers (Figure 4.6B–D).

Organic Rocks

Most organic rocks are composed of fragments of calcite shells of invertebrate animals. Thus, they form varieties of limestone, such as coquina, skeletal limestone, and chalk (Figure 4.5D–F). Peat and coal (Figure 4.6E–F) are sedimentary rocks formed from alteration of plant debris.

	TEXTURE		COMPOSITION	ROCK NAME
CLASTIC	Coarse grained		Rounded fragments of any rock type—quartz, quartzite, and chert dominant	**CONGLOMERATE**
			Angular fragments of any rock type—quartz, quartzite, and chert dominant	**BRECCIA**
	Coarse to fine grained		Poorly sorted, nonstratified, angular fragments of any rock type	**TILLITE**
	Medium grained		Quartz and rock fragments	**SANDSTONE**
	Fine grained		Quartz and clay minerals	**SILTSTONE**
	Very fine grained		Quartz and clay minerals	**SHALE**
CHEMICAL OR ORGANIC	Medium to coarse grained			**CRYSTALLINE LIMESTONE**
	Microcrystalline, conchoidal fracture			**MICRITE**
	Aggregates of oolites			**OOLITIC LIMESTONE**
	Fossils and fossil fragments loosely cemented	Calcite ($CaCO_3$)		**COQUINA**
	Abundant fossils in calcareous matrix			**FOSSILIFEROUS LIMESTONE**
	Shells of microscopic organisms and clay—soft			**CHALK**
	Banded calcite			**TRAVERTINE**
	Textural varieties similar to limestone	Dolomite ($CaMg[CO_3]_2$)		**DOLOMITE**
	Cryptocrystalline, dense, conchoidal fracture	Chalcedony (SiO_2)		**CHERT**
	Fine to coarse crystalline	Gypsum ($CaSO_4 \cdot 2H_2O$)		**ROCK GYPSUM**
	Fine to coarse crystalline	Halite ($NaCl$)		**ROCK SALT**
	Fibrous	Brownish plant material—soft, porous		**PEAT**
	Dense	Highly altered plant remains—carbon		**COAL**

FIGURE 4.3
Classification of Sedimentary Rocks

A. Conglomerate consists of coarse fragments (greater than 2 mm in diameter) held together by a matrix of sand, clay, and cement. The individual pebbles are usually well rounded and moderately well sorted. Cobbles and pebbles in conglomerate may consist of any mineral or rock, but resistant materials such as quartz, quartzite, and chert are especially common. Fragments of limestone, granite, or other rock types may predominate, however, in some deposits.

The only agents capable of transporting the large fragments that make up a conglomerate are swiftly moving water, glaciers, and the direct action of gravity. The most important environments for conglomerate deposition are alluvial fans, river channels, and beaches.

Quartz pebble
Quartz sand

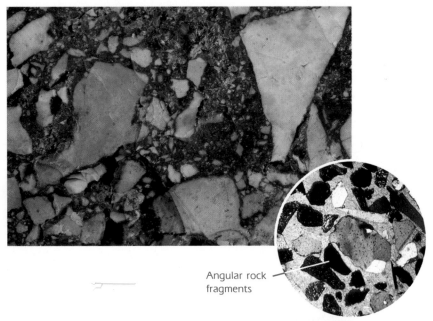

B. Breccia is a coarse-grained clastic rock in which fragments are angular and show little evidence of abrasion. Commonly, material is poorly sorted, with a fine-grained matrix. The most significant breccia deposits result from meteorite impact, landslides, and other types of mass movement. Movement along a fault also produces a breccia zone, as does a collapsed cave.

Angular rock fragments

C. Tillite is composed of poorly sorted, unstratified sediment deposited by glaciers. The particles are typically angular. In many deposits, fine-grained sand, silt, and clay-size materials dominate.

Clay & silt
Angular fragment

FIGURE 4.4
Clastic Rocks

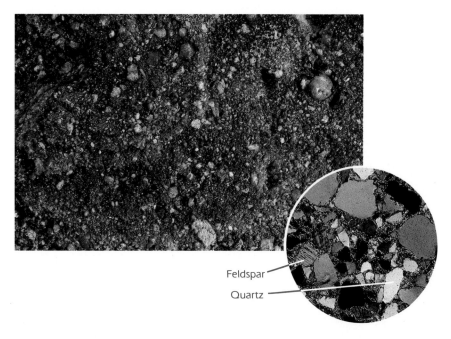

Feldspar
Quartz

D. Arkose is a clastic rock composed of at least 25% feldspar. Quartz is the other major constituent. In most arkoses, the grains are in the sand-size range, but some specimens are coarse enough to be termed conglomerates. In this specimen, feldspar is pink, and close inspection reveals its abundance. In thin section, feldspar grains commonly have a plaid pattern and are nearly as abundant as quartz, which appears gray-blue. Calcite cement appears brown in this thin section. Arkoses form by rapid erosion of a granitic terrain and are commonly deposited in alluvial fans.

Quartz

E. Sandstone is a clastic sedimentary rock consisting mostly of grains ranging from 1/16 to 2 mm in diameter. Grains are generally rounded and show the effects of abrasion, as seen in this thin section. Quartz is usually the dominant mineral, although rock fragments, feldspar, garnet, mica, and other minerals may be present in varying amounts. Calcite, quartz, and iron oxide are the main cementing materials. Sandstones are generally stratified, and are colored buff, red, brown, yellow, or green, depending on impurities and cementing agents. Sandstones may accumulate in a wide variety of environments, such as beaches, deserts, floodplains, and deltas.

Clay

F. Shale is a fine-grained clastic rock consisting of particles less than 1/256 mm in diameter. Most shales are characteristically laminated or thin bedded, and the structure is usually well expressed in a hand specimen. Quartz, mica, and clay minerals are dominant constituents. Individual particles are too small to be seen without high magnification. Calcite may be present, usually as cement, in amounts ranging up to 50%.

Eighty percent of exposed sedimentary rocks are shale. The particles accumulate in quiet-water environments, such as shallow-marine waters, lagoons, floodplains, and lakes.

47

Calcite

A. Crystalline Limestone (Sparite) is composed of interlocking calcite crystals large enough to be seen with the unaided eye or with low-power magnification. Well-preserved fossils may be common, and bedding planes are generally distinct. Good rhombohedral cleavage in the calcite crystals is commonly well expressed in a microscopic view. Crystalline limestone is typically formed in shallow seas.

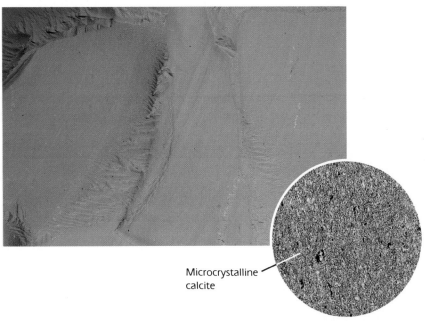

Microcrystalline calcite

B. Microcrystalline Limestone (Micrite) is composed almost entirely of microscopic calcite crystals so small they are difficult to discern even under the microscope. The rock has uniform texture, is very dense, and characteristically breaks with a conchoidal fracture. It is commonly buff or light yellow, but may range in color to dark gray or black if organic material is abundant. Microcrystalline limestone forms in warm, quiet, shallow seas.

Oolites

C. Oolitic Limestone is composed of small spherical grains, each of which has a concentric layered structure similar to that of an onion. The grains range up to 2 mm in diameter and consist of concentric layers formed around a grain nucleus of fine silt, clay, or shell fragments. Calcite commonly cements the spherical particles together to form a rock in which the texture closely resembles the well-rounded grains in a quartz sandstone. Stratification, sorting, cross-bedding, and other sedimentary features characteristic of sandstone are also common in oolitic limestone.

Oolites originate in shallow water where currents and waves agitate particles of silt and small shell fragments. These act as nuclei and pick up calcite, which adheres to the grain surface much like a rolling snowball.

FIGURE 4.5
Limestones

Shell
fragments

D. Coquina is the name given to a weak, porous, poorly cemented limestone composed almost exclusively of shells and shell fragments. The shell material in a coquina is commonly broken into platy fragments, which are rounded, abraded, and moderately well sorted.

Coquina is thus a particular type of clastic rock composed of fragmental materials that originated in the sedimentary environment, rather than having been transported from an eroded landmass. Coquinas are forming today along the southern Atlantic coast and in the Bahama Islands, where shell material is washed on shore and accumulates as calcareous sand fragments on a beach or in shallow water.

Shell
fragments

E. Skeletal (Fossiliferous) Limestone is composed primarily of hard parts of invertebrate organisms such as mollusks, corals, and crinoids. This material is commonly cemented with microcrystalline calcite and may form a dense rock. Fossils often weather out in relief and show the rock texture and structure. In thin section, details of fossil shell structure are usually well expressed. Fossiliferous limestone forms in warm, shallow seas where marine life is abundant.

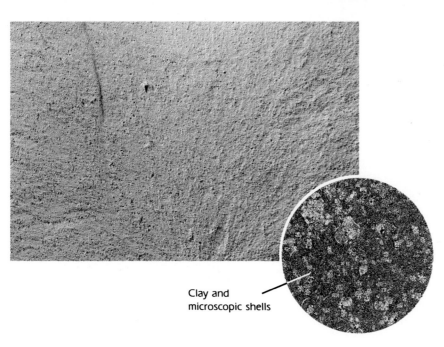

Clay and
microscopic shells

F. Chalk is a soft, porous, fine-textured limestone composed of shells of microscopic organisms, mostly foraminifera. It may also contain varying amounts of mud. Coloration is commonly white or buff. The best-known chalks are of Cretaceous age and include the famous white chalk cliffs on both sides of the English Channel, the Selma chalk of the Gulf Coast, and the Niobrara chalk of Kansas. Chalk is considered a shallow-water deposit, formed by the accumulation of the shells of floating organisms.

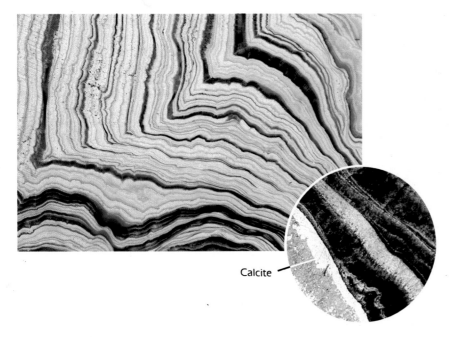

A. Travertine is a calcium carbonate (limestone) deposit formed in caves and springs. It is characteristically banded with alternating light and dark layers. The layers are created by small amounts of iron oxide that accumulate during successive periods of deposition. The well-known flowstone, dripstone, stalagmite, and stalactite deposits in caves are all varieties of travertine. Most travertine consists of recent, relatively small deposits. It is of no great geologic importance, but its beauty and variety of form make it memorable.

Calcite

B. Rock Gypsum is a chemical precipitate composed almost exclusively of aggregates of the mineral gypsum. It is commonly white or delicately colored in various shades of orange or light red due to small amounts of iron oxide. One of its distinctive characteristics is that it can be scratched with a fingernail. Gypsum is commonly massive, but in some deposits, thin, delicate laminae form as a result of seasonal influxes of clay. Gypsum originates from the evaporation of saline lakes or seawater trapped in restricted bays. The presence of this mineral thus indicates an arid or semiarid climate at the time of formation.

Gypsum

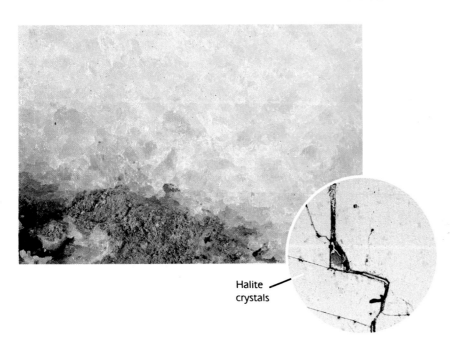

C. Rock Salt is composed of the mineral halite. Crystals may be fine, medium, or coarse and are generally colorless or stained various shades of red by iron oxide or red clay. Rock salt is an evaporite deposit and originates in a manner similar to gypsum. Thick salt deposits are found in Michigan, New York, and Kansas, and in relatively recent deposits from saline lakes in the arid western states. An interesting and important characteristic of salt is that it flows at relatively low temperatures and pressures. Salt from deeply buried strata may rise as piercement plugs or salt domes.

Halite crystals

FIGURE 4.6
Other Sedimentary Rocks

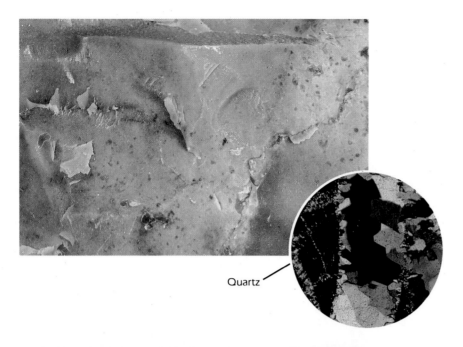

Quartz

D. Chert is a common chemical rock, composed of microcrystalline quartz. Coloration may be white or various shades of gray, green, blue, red, yellow, or black. Under high magnification, a fibrous or granular texture can be seen. Chert commonly occurs as nodules in limestone. Like other varieties of quartz, such as flint and opal, chert is characterized by a hardness of 7 and conchoidal fracture. Migrating groundwater may deposit large amounts of chert in the pore spaces of limestone.

Plant fragments

E. Peat is a dark-brown or black organic deposit produced by compaction and the partial decomposition and disintegration of plants. It forms in bogs and swamps and represents the initial stage in the formation of coal. Organic matter constitutes from 70–90% of the total accumulation. Mineral compounds are an insignificant component. Peat formation requires rapid plant growth and reproduction and a minimum of microorganism activity. In some swamps, peat deposits may be tens of feet thick and cover many square kilometers.

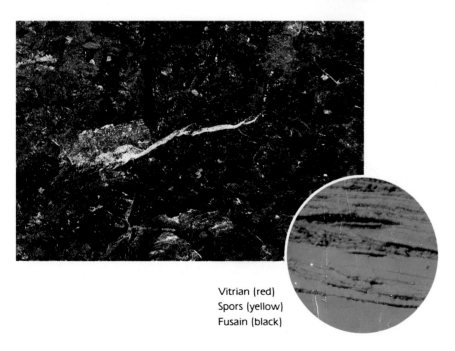

Vitrian (red)
Spors (yellow)
Fusain (black)

F. Coal is composed of highly altered plant remains and various amounts of clay. It is opaque and noncrystalline, with coloration ranging from light brown to black. Coalification results from the burial and compaction of peat. Coal is classified according to the degree of change: Lignite (brown coal) forms first; deeper burial and greater compaction produce bituminous (soft) coal; and prolonged burial and pressure produce anthracite (hard) coal.

Origin of Sedimentary Rocks

Sedimentary rocks are derived from the weathering and erosional debris of other rocks. This material is deposited at Earth's surface, under conditions of normal temperature and pressure. The major processes involved in the formation of sedimentary rocks are (1) physical and chemical weathering of the parent rock material; (2) transportation of these weathered products by running water, wind, gravity, or ice; (3) deposition of the material in a sedimentary basin; and (4) compaction and cementation of sediment into solid rock. These processes operate at or near Earth's surface and are therefore observable.

A significant result of the transportation and deposition of sediment is the sorting and segregation of the parent rock material into deposits of similar grain size and composition. Materials similar in size and weight are washed and winnowed by wind and water, and deposited in a sedimentary environment according to the level of mechanical energy operating at the time of deposition. For example, coarse gravels may be deposited in an alluvial fan near a mountain front or on a river bar, whereas sand is transported downstream to the sea to be concentrated on a beach. Currents and wave action wash away the finer material, which is deposited ultimately in an environment of low mechanical energy, such as a marsh or lagoon. Chemical sorting occurs contemporaneously; the more soluble materials dissolve and are removed in solution and may later be precipitated in a lake or an ocean basin.

The term *sedimentary environment* refers to the place where the sediment is deposited and to the physical, chemical, and biological conditions that exist there. The diagram to the right shows, in a general way, the regional settings of some major sedimentary environments. The description of each environment summarizes some of the important characteristics of the sedimentary rocks formed in each environment.

Alluvial Fans

Alluvial fans are stream deposits that accumulate near a mountain front in a dry basin. They typically contain poorly sorted coarse gravel and boulders. Fine-grained sand and silt may be deposited near the margins of the fan.

Eolian Environments

Wind is an effective sorting agent and will selectively transport sand and dust. It leaves gravel behind but moves sand near the surface and lifts dust-sized particles high in the atmosphere and transports them thousands of miles before they accumulate as a thin blanket of loess. Windblown sand commonly accumulates in dunes, characterized by well-sorted, fine grains. The dominant sedimentary structure in eolian environments is large-scale cross-bedding.

Shoreline Environments

Beaches, bars, and spits commonly develop along low coasts and partly enclose quiet-water lagoons and tidal flats. The sediment in these environments is well washed by wave action and is typically clean, well-sorted quartz sand. Behind the bars, and adjacent to the beaches, fine silt and mud are often deposited as tidal flats.

Glacial Environments

Glaciers transport but do not effectively sort material. The resulting deposit is an unsorted, unstratified accumulation of angular boulders, gravel, sand, and fine silt.

Floodplains

The great rivers of the world typically meander across a flat floodplain before reaching the sea; over time, a considerable amount of sediment is deposited on these plains. Rocks formed in a floodplain environment are commonly channels of sandstone, deposited on the point bar of a meander and enclosed in a shale deposited in the backswamps.

Deltas

A delta is a large accumulation of sediment that is deposited at the mouth of a river. A delta is one of the most significant sedimentary environments and includes a number of subenvironments, such as stream channels, floodplains, beaches, bars, and tidal flats. The deltaic deposit as a whole consists of a thick accumulation of silt, mud, and sand.

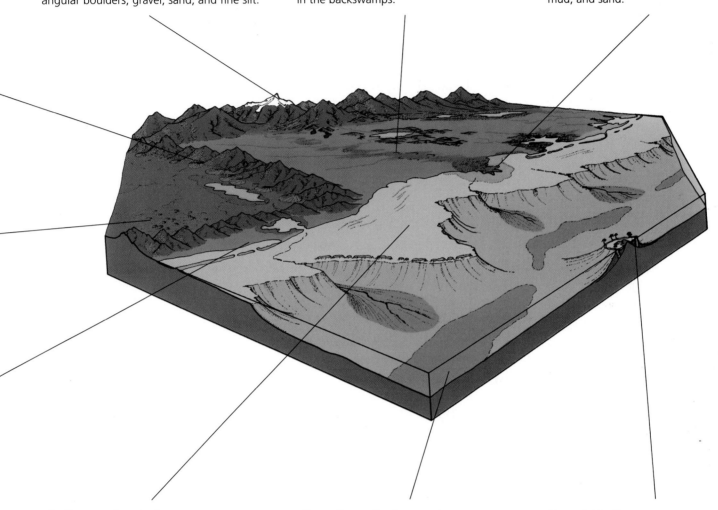

Shallow-Marine Environments

Shallow seas are widespread along the continental margins; in the past, they were even more extensive than today. Sediments deposited in shallow-marine waters form extensive layers of well-sorted sand, shale, and limestone that typically occur in a cyclical sequence as a result of shifting environments from changes in sea level.

Deep-Ocean Environments

The deep ocean adjacent to the continents receives a considerable amount of sediment transported from the continental margins by turbidity currents. As a current moves across the deep-ocean floor, its velocity decreases gradually, and the sediment carried in suspension settles. The resulting deposit is a wide-spread sediment layer in which the grain size grades from coarse, at the base of the layer, to fine, at the top. Deep-sea deposits are thus characterized by a sequence of graded beds.

Organic Reefs

An organic reef is a solid structure composed of the shells and secretions of marine organisms. The reef framework is typically built by corals and algae, but many other types of organisms contribute to the reef community. Together, these organisms produce a highly fossiliferous limestone.

PROBLEMS

1. Identify the specimens provided by your instructor. To do this efficiently, we suggest the following procedures:

    a. Determine whether the rock is composed of calcium carbonate. (Rocks composed mostly of $CaCO_3$ will effervesce in dilute HCl; others will not.)

    b. Determine whether the rock has a clastic (fragmental) or a nonclastic texture. (Use a hand lens or a microscope if one is available.)

    c. Determine the grain size and matrix:

- Clastic—Decide whether the grains are predominantly gravel, sand, silt, or clay size. Note the abundance of matrix material.

- Crystalline carbonates—Decide whether the texture is coarse, medium, or fine crystalline.

- "Clastic" carbonates—Decide whether the texture is predominantly oolitic or fossiliferous.

- Determine the composition of the matrix.

- With the above information, refer to Figure 4.3 to find the rock name.

2. Compare each specimen with the illustrations and descriptions on pages 46–51. How does your specimen differ from the illustration of the same rock type?

3. Are the differences of fundamental importance (differences of composition or texture), or are they minor (a difference in color or overall shape)?

4. Study the photographs of hand specimens and photomicrographs on pages 55–56. Using the information on composition, determine the name of each rock and describe its origin.

5. List the environments in which shale accumulates. What does each of these have in common with the others?

6. What evidence indicates that sandstone forms at Earth's surface?

7. How is $CaCO_3$ extracted from seawater?

8. What is the origin of clay minerals?

9. What rock types commonly occur as fragments in sandstone?

10. Why is olivine a rare mineral in sandstone—even on the beaches of volcanic islands?

11. What properties are common to all limestones, regardless of texture?

12. Many limestones are dense, fine grained, and black. How can you distinguish between a limestone and a basalt?

13. How do oolites originate? Where are they deposited?

14. How are rock salt and rock gypsum formed?

15. What is an arkose? How does it differ from a quartz sandstone?

16. What is the major source rock of an arkose?

17. Study the environments shown on page 53 and list the rock types likely to be formed in each major environment.

18. Why does sandstone form in so many different environments?

19. What geologic events are indicated by a clean, well-sorted quartz sandstone?

20. How would these events differ in the formation of an arkose?

21. Many limestones have a crystalline texture. What features of a limestone indicate that it is not an igneous rock?

22. What are the major processes involved in the genesis of sedimentary rocks?

23. What is the difference between a tuff and a quartz sandstone?

24. Many beaches on the volcanic islands in the Pacific Ocean are white. The islands, however, are composed of black basalt, which has no quartz. What is the composition of the beach sand? How did it originate?

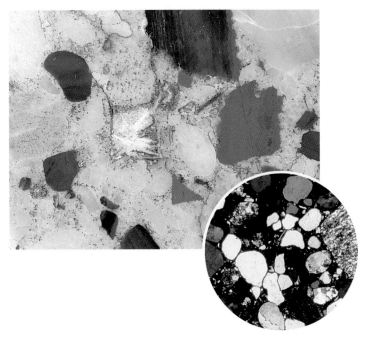

Specimen A

Composition
Jasper	30%
Quartzite	50%
Quartz	18%

Rock name _____

Origin_____

Specimen B

Composition
$CaSO_4$	95%
Iron	2%
Clay	3%

Rock name _____

Origin_____

Specimen C

Composition
Feldspar	20%
Quartz	70%
Rock fragments	5%
Clay	5%

Rock name _____

Origin_____

Specimen D

Composition
Quartz   98%

Rock name _____

Origin_____

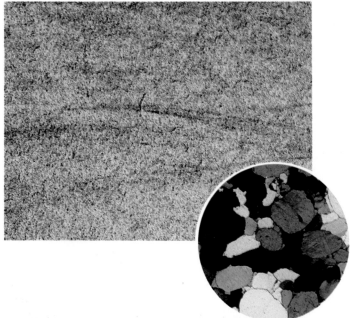

Specimen E

Composition
Quartz   95%

Rock name _____

Origin_____

Specimen F

Composition
$CaCO_3$ 97%
Clay   3%

Rock name _____

Origin_____

METAMORPHIC ROCKS

■ OBJECTIVE

To recognize the major metamorphic rock types and to understand the significance of their texture and composition.

■ MAIN CONCEPT

Metamorphic rocks are classified on the basis of texture and composition. Two main groups are recognized: (1) foliated and (2) nonfoliated. Regional metamorphism occurs in the roots of mountain belts formed at convergent plate margins. Contact metamorphism develops at the margins of igneous intrusions.

■ SUPPORTING IDEAS

1. Foliation results from recrystallization and the growth of new minerals.
2. The three main types of foliation are (a) slaty cleavage, (b) schistosity, and (c) gneissic layering.
3. Nonfoliated texture develops by the recrystallization of rocks composed predominantly of one mineral, such as sandstone (quartz) or limestone (calcite).

Metamorphic rocks are rocks that have undergone a fundamental change as a result of heat, pressure, and the chemical action of pore fluids and gases. The parent rock material, before metamorphism, might include marine sediments that formed at surface temperatures and pressures, interbedded lava that crystallized at high temperatures, and granite that subsequently intruded the entire sequence. These materials would be unstable at the prevailing temperatures and pressures in the crustal interior, especially in a mountain belt formed by the collision of lithospheric plates. Under conditions of increased temperature or pressure or both, the parent rock reacts in such a way that a new assemblage of minerals is produced—an assemblage that is stable under the new conditions of higher temperature and pressure. Metamorphic changes always tend to establish equilibrium between the rock material and the prevailing physical and chemical environment.

The effects of metamorphism include (1) chemical recombination and the growth of new minerals, with or without the addition of new elements from circulating liquids and gases; (2) deformation and rotation of the constituent mineral grains; and (3) the recrystallization of minerals to form

larger grains. The net result is rock of greater crystallinity, increased hardness, and new structural features that commonly exhibit the effects of flow or other expressions of deformation. All three major rock types can be metamorphosed, but intrusive igneous rocks and previously metamorphosed rocks are affected less by metamorphic processes than are the sedimentary rocks that developed at Earth's surface.

METAMORPHIC TEXTURE

Foliated Textures

Foliation is a planar element in metamorphic rocks. It may be expressed (1) by closely spaced fractures (slaty cleavage), (2) by the parallel arrangement of platy minerals (schistosity), or (3) by alternating layers of different mineral composition (gneissic layering). Foliation is usually developed during metamorphism by directed stresses that cause differential movement or recrystallization. It is a fundamental characteristic of metamorphic rocks and is a basic criterion of the classification system. The various types of foliation are shown in Figure 5.1A–D.

Slaty Cleavage. Slaty cleavage is a type of foliation expressed by the tendency of a rock to split along parallel planes. Do not confuse this planar feature with bedding planes, which are a sedimentary structure. Slaty cleavage results from the parallel orientation of microscopic platy minerals such as mica, talc, or chlorite. In a metamorphic environment, these minerals grow with flat surfaces perpendicular to the applied forces. The perfect cleavage within each tiny mineral grain is thus oriented in the same direction. This creates definite planes of weakness throughout the entire rock and causes it to break along nearly parallel planes. Such a surface is shown in Figure 5.1A. Note that slaty cleavage commonly cuts across bedding planes, the thin light and dark lines in the specimen in the figure. It is the product of a relatively low intensity of metamorphism.

Schistosity. Schistosity develops from more intense metamorphism. Mica, chlorite, and talc form larger, visible crystals and, as a result, the rock develops a distinctly planar element (Figure 5.1B). Schistosity is thus similar to slaty cleavage, but the platy mineral crystals are much larger, and the entire rock appears coarse grained. The

A. Slaty Cleavage is a type of foliation expressed by the tendency of the rock to split into thin layers. It results from the parallel orientation of microscopic grains of mica, chlorite, or other platy minerals. This photo shows the surface of a cleavage plane. Relic bedding can be seen as faint, reddish layers.

B. Schistosity is a type of foliation resulting from the parallel to sub-parallel orientation of large, platy minerals such as mica, chlorite, and talc. This photo shows the surfaces of the planes of schistosity. A section through the schistosity plane is shown in the specimen's left margin. The small, dark grains are garnet.

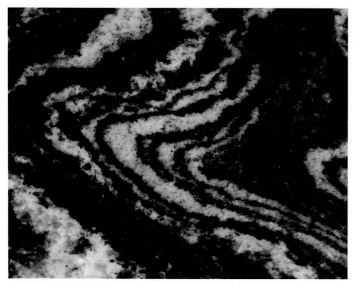

C. Gneissic Layering is a coarse-grained, foliated texture in which the minerals are segregated into layers that can be several centimeters thick. Light-colored layers commonly contain quartz and feldspar; dark-colored layers are commonly composed of biotite and amphibole.

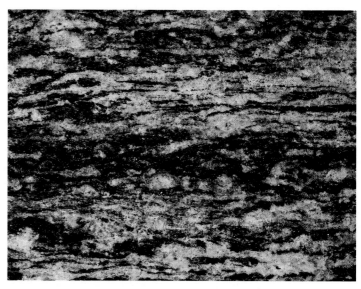

D. Gneissic Layering in some rocks may be thin, discontinuous lenses that impart a distinctive fabric to the rock.

FIGURE 5.1
Foliated Textures

increase in crystal size represents a higher grade of metamorphism in which garnet, amphibole, and other nonplaty minerals also develop. Schistosity typically occurs in mica-rich rocks.

Gneissic Layering. Gneissic layering is a type of foliation in which the planar element is produced by alternating layers of different mineral composition. Rocks with gneissic layering are characteristically coarse grained and represent a higher grade of metamorphism in which the minerals are recrystallized, stretched, crushed, and rearranged completely. Feldspar and quartz commonly form light-colored layers that alternate with dark layers of ferromagnesian minerals (Figure 5.1C).

A. Stretched Pebbles result from the metamorphism of conglomerate and impart a distinctive linear fabric to the rock. Strong foliation does not develop. The polished surface of this specimen is a cross section, oriented diagonally across the long axis of the stretched pebbles. The actual length of the pebbles shown here is nearly 10 cm.

B. Recrystallization results from metamorphism and may develop a massive, nonfoliated texture, as shown here. Calcite crystals have grown much larger than the original grains but are essentially equidimensional. The texture is therefore granular and coarsely crystalline, but nonfoliated.

C. Fused Sandstone Grains produce a dense, massive metamorphic texture without foliation. Grains are commonly stretched and deformed and interlock to form a dense mass.

FIGURE 5.2
Nonfoliated Textures

Nonfoliated Textures

Some metamorphic rocks do not possess foliation but appear massive and structureless, except for elongated grains or other linear features resulting from directional stress. Examples of nonfoliated texture are shown in Figure 5.2A–D. Nonfoliated rocks are commonly formed from a parent rock composed largely of a single mineral, such as sandstone or limestone.

Lineation, such as the elongated pebbles in Figure 5.2A, may develop in nonfoliated rocks. The pebbles were originally round but have been stretched by the directed stress within the rock. Sand grains in a metamorphosed sandstone often show similar expressions of deformation, or the grains may be fused into a dense, compact mass of interlocking particles. Deformation of limestone will produce streaks of material that contains large amounts of organic matter.

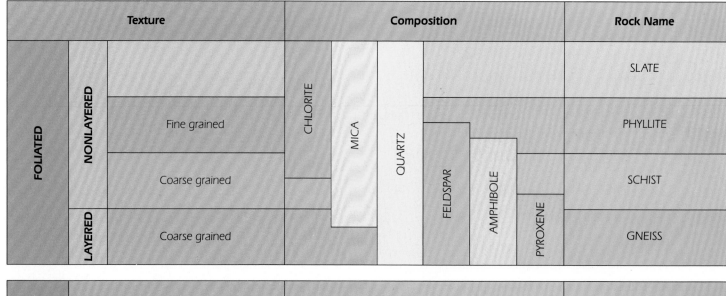

Texture			Composition	Rock Name
FOLIATED	NONLAYERED		CHLORITE / MICA / QUARTZ / FELDSPAR / AMPHIBOLE / PYROXENE	SLATE
		Fine grained		PHYLLITE
		Coarse grained		SCHIST
	LAYERED	Coarse grained		GNEISS

Texture	Composition	Rock Name
NONFOLIATED — Coarse grained	Deformed grains of any rock type	METACONGLOMERATE
NONFOLIATED — Fine to coarse grained	Quartz	QUARTZITE
	Calcite or dolomite	MARBLE

FIGURE 5.3
Classification of Metamorphic Rocks

CLASSIFICATION OF METAMORPHIC ROCKS

The formation of metamorphic rocks is so complex that developing a satisfactory classification system is difficult. The most convenient scheme is to group metamorphic rocks by structural feature, with further subdivision based on composition. Using this classification, two major groups of metamorphic rocks are recognized: (1) those that are *foliated* (possess a definite planar structure), and (2) those that are *nonfoliated,* that is, massive and structureless. The foliated rocks can then be subdivided further, according to the type of foliation. Finally, a large variety of rock types can be recognized in each group, according to the dominant minerals. The basic framework for this classification system is shown in Figure 5.3. Examples of the major types of metamorphic rocks are shown in Figures 5.4 and 5.5. They are shown in the same order in which they are listed in Figure 5.3.

Serpentinite is a fine-grained metamorphic rock that differs from the more conventional metamorphic rocks shown in the classification chart above. It may be massive or it may be weakly foliated with weak schistosity.

Serpentinite is derived from seawater circulation through oceanic crust near the oceanic ridge. Large volumes of cold seawater seep downward through fractures and fissures in the pillow basalts and sheeted dikes and are heated and then rise, altering the surrounding basalt. Serpentinite is thus one of the most abundant metamorphic rocks on Earth although most is hidden beneath the ocean.

PHOTOGRAPHS OF ROCK SPECIMENS

The photographs in Figures 5.4 and 5.5 illustrate the characteristics of the major types of metamorphic rocks, as seen in hand specimens and under a microscope. The hand specimens are actual size and the photomicrographs show a magnification of approximately 20 times. Some specimens have a polished surface, which better shows their textual features. Others have a natural fractured surface. *Remember, the photographs are not a key to rock identification!* Use them as a visual reference only.

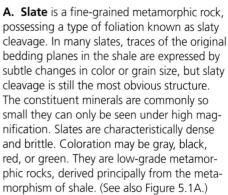

A. Slate is a fine-grained metamorphic rock, possessing a type of foliation known as slaty cleavage. In many slates, traces of the original bedding planes in the shale are expressed by subtle changes in color or grain size, but slaty cleavage is still the most obvious structure. The constituent minerals are commonly so small they can only be seen under high magnification. Slates are characteristically dense and brittle. Coloration may be gray, black, red, or green. They are low-grade metamorphic rocks, derived principally from the metamorphism of shale. (See also Figure 5.1A.)

Phyllite is similar to slate but is distinguished by a satin luster, or sheen, developed on the planes of foliation. The luster results from the growth of larger mica crystals commonly found in slate.

Mica

B. Schist is a metamorphic rock in which foliation is due to the parallel arrangement of relatively large crystals of platy minerals. Muscovite, biotite, chlorite, and talc are the important platy constituents. Feldspars are rare, but quartz, garnet, and hornblende are common accessory minerals.

Schists may be further classified on the basis of the more important minerals present. The most common types are chlorite schists, muscovite schists, hornblende mica schists, and garnetiferous mica schists. The unifying characteristic, regardless of composition, is that foliation results from the parallel arrangement of relatively large, platy minerals. Schists are produced by a metamorphism of higher intensity than that which produces slates. Under such conditions, a variety of parent-rock types—including basalt, granite, sandstone, and shale—may be converted to schists.

Mica

C. Blueschist results from the high-pressure, low-temperature metamorphism of oceanic crust (basalt and oceanic sediment) in the subduction zone below a trench. At a subduction zone, the rapid and deep subsidence of the cool lithospheric slab creates the high pressure. The low temperature persists because the descending plate heats up slowly, while the pressure increases rapidly.

Mica
Quartz

FIGURE 5.4
Foliated Rocks

D. Greenschist results from the high-pressure, high-temperature metamorphism of a variety of rock types. It forms in the roots of mountain belts, where high pressure is created at depths similar to those at which blueschist is formed. In the mountain roots, however, high temperature also occurs as a result of the ascending magma generated by melting at convergent plate margins. Under some high-temperature conditions, blueschist is metamorphosed into greenschist.

Mica
Quartz

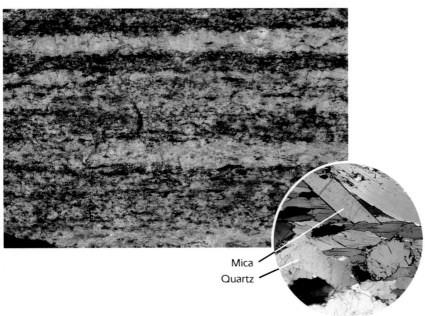

E. Gneiss is a metamorphic rock in which foliation results from layers composed of different mineral groups. Feldspar and quartz are the chief minerals in gneiss, with significant amounts of mica, amphibole, and other ferromagnesian minerals. Gneiss has a coarse-grained, granular texture. It resembles granite in composition but is distinguished from it by the foliation. (See also Figures 5.1C and D.)

Mica
Quartz

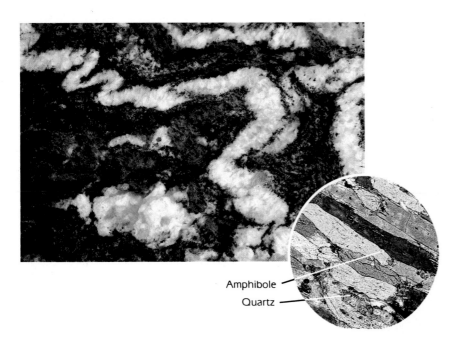

F. Gneiss may be foliated, with semicontinuous layers of light and dark minerals, as shown in Figure 5.4E (above), or with highly contorted, well-defined layers like those shown here. In many rock bodies, layers expressing foliation are several inches thick. Gneisses are among the most abundant metamorphic rocks. They represent a high grade of metamorphism and may originate from various granite-rhyolite rocks or from lower-grade metamorphic and sedimentary rocks.

Amphibole
Quartz

FIGURE 5.4 *(Continued)*

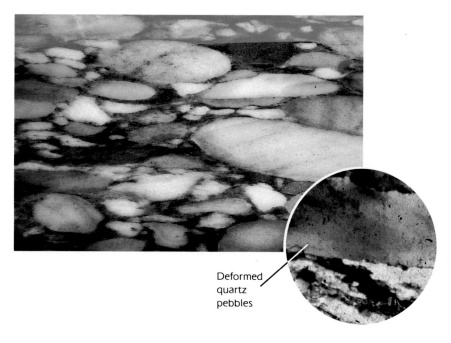

Deformed
quartz
pebbles

A. Metaconglomerate is conglomerate that has been altered by heat and pressure to such an extent that individual pebbles are stretched, deformed, and fused. Stretched pebbles commonly show a definite lineation, related to the orientation of stresses, but the rock is not strongly foliated. Most metaconglomerate is so indurated that the rock fractures across the pebbles as easily as around them. In this specimen, the matrix of sand grains has been squeezed and fused to the larger pebbles. As a result, the pebble borders appear in the photograph to be indistinct or out of focus. Under a microscope, the sand-silt matrix will show deformation, similar to that exhibited by the pebbles. This photograph is a polished surface cutting across pebbles that are up to 20 cm long.

Quartz

B. Quartzite is a nonfoliated metamorphic rock, composed principally of quartz. In some deposits, quartz is the only significant mineral present. The individual grains of quartzite are stretched, deformed, interlocked, and fused, so the rock breaks across the grains indiscriminately. Pure quartzite is derived from quartz sandstone, but some quartzites may contain as much as 40% of other minerals—mica being one of the most abundant. Although quartzite is nonfoliated, some formations that were originally interbedded with shale contain incipient slaty cleavage or relict bedding, which imparts a planar element to the rock.

Calcite

C. Marble is a nonfoliated metamorphic rock, composed principally of calcite or dolomite. The crystals are commonly large and interlock to form a dense crystalline structure. Bands or streaks of organic impurities, resulting from flowage or extreme deformation, are common in some deposits. Coloration may be white, pink, blue-gray, or brown. Like limestone, marble is characterized by its softness and its effervescence with hydrochloric acid.

FIGURE 5.5
Nonfoliated Rocks

ORIGIN OF METAMORPHIC ROCKS

Metamorphism

Regional Metamorphism. Regional metamorphism results from the combined action of high temperatures and directed stresses. It is common where lithospheric plates collide and the rocks are folded, crushed, sheared, and stretched. Rocks in the roots of mountain belts are generally altered by recrystallization, growth of new minerals, and reorientation of existing minerals.

Careful field and laboratory studies indicate that certain mineral assemblages are indicative of varying degrees of metamorphism. Chlorite, talc, and mica, for example, form under relatively low temperatures and pressures, whereas garnet and amphibole develop under more intense conditions.

Contact Metamorphism. Contact metamorphism results from high temperatures and vigorous solution activity, concentrated near the contacts of a cooling magma. This type of metamorphism is localized and diminishes rapidly away from the intrusive body. Heat and the chemical activity of fluids associated with the magma are the major metamorphic agents. The affected rocks generally recrystallize to form hard, massive bodies.

Metamorphism and Tectonics

Metamorphism at Convergent Plate Margins. Most metamorphic rocks on continents were formed in mountain belts at convergent plate margins. These rocks represent an important phase in the growth of continents by accretion (Figure 5.6A). The major steps in the process of continental accretion can be summarized as follows.

Sediment derived from the erosion of a continent (or an island arc) is transported and deposited along the continent's trailing margin. Convecting currents in the asthenosphere may change the direction of plate movement, and ultimately the trailing margin, with the thick deposit of sediments, collides with another plate. The sediments are thus deformed into a mountain range. Rocks in the roots of the mountain are subjected to high pressure because of deep burial and to high temperatures because of the generation and migration of magma that is intruded into the mountain belt. The deep burial and the tremendous horizontal pressures of the converging plates cause the rocks in the roots of the mountain belt to partially melt and recrystallize. New minerals grow under the horizontally directed stresses and typically develop a foliation that is perpendicular to the stress (essentially vertical). High-grade metamorphism of the sediments forms gneisses and schists, deep in the mountain roots. Slates develop at shallower depths. Blueschists result from the low-temperature, high-pressure metamorphism of deep-marine shale deposited on the continental margins. Greenschists form in the zone of deep burial as a result of high pressure and high temperature. Some completely melted rock material may rise to the surface and cool as dikes, injected between the planes of foliation. Erosion of a mountain range causes the mountain belt to rise due to isostatic adjustment. Ultimately, the metamorphic rock, formed deep in the mountain root, is exposed at the surface and forms a new segment of the continental shield.

Metamorphism at Divergent Plate Margins. Another type of metamorphism occurs at the oceanic ridge and is completely different from that produced by mountain building. Here, metamorphism is produced by the chemical action of hot seawater with little or no intense heat and pressure. At the oceanic ridge, the major tectonic stress is extension that creates numerous open fissures through which seawater can seep downward to depths of 2 to 3 km near the base of the dike complex. As the water circulates through the hot volcanic rocks, it is heated to temperatures as high as 450 °C. This hot, saline water reacts with the olivine, pyroxene, and plagioclase in the basalt to form new minerals—chlorite, sodium-rich plagioclase, talc, and serpentinite.

The circulation of hot water occurs along the entire oceanic ridge and represents a fundamental geologic process. The total amount of water circulating through the oceanic crust each year is equivalent to 2% of the annual discharge of all rivers on the planet—enough to recycle the entire volume of the oceans every 5 to 10 million years.

PROBLEMS

1. Identify the rock specimens provided by your instructor. To do this efficiently, follow these procedures:

    a. Determine whether the rock is foliated or nonfoliated.

    b. If the rock is foliated, determine the type of foliation.

    c. Identify the minerals present.

    d. With the preceding information, refer to Figure 5.3 and identify the rock by name.

    e. On the basis of texture and composition, determine, if possible, the original rock type.

    f. What is the difference between your specimens and the photographs of the same rock types shown in this manual?

2. Examine the photographs of the rocks shown on pages 66–67. Using the information about composition, determine the name of each rock and describe its origin.

3. What are the major differences between regional and contact metamorphism?

4. How is the orientation of foliation related to stress?

5. Which metamorphic rocks are likely to contain garnet?

6. Why do clay minerals change to mica in the metamorphism of shale?

7. Is contact metamorphism likely to produce foliation? Explain your answer.

8. How can you distinguish between the following rock types?

    a. Marble and quartzite

    b. Slate and shale

    c. Conglomerate and metaconglomerate

    d. Granite and gneiss

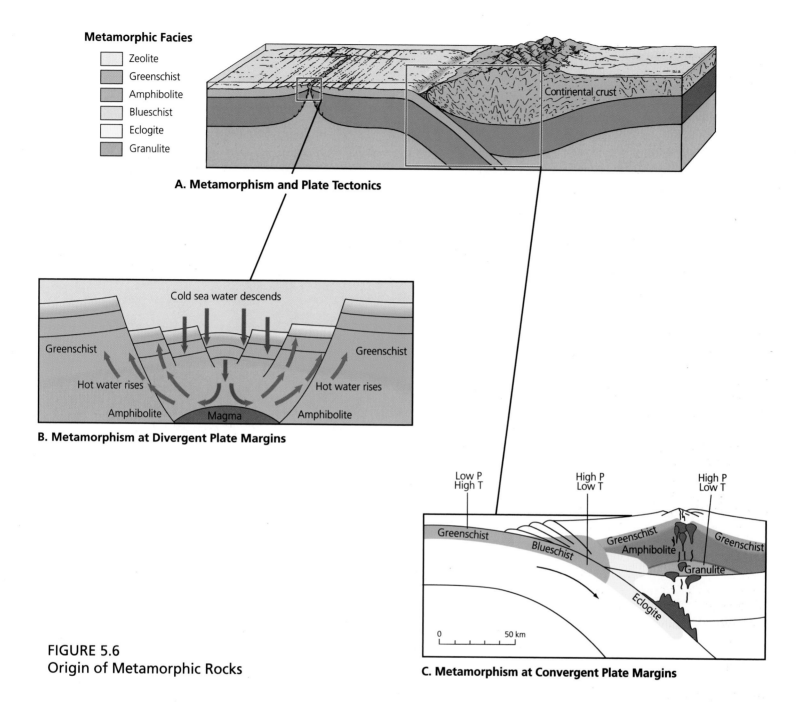

Metamorphic Facies

- Zeolite
- Greenschist
- Amphibolite
- Blueschist
- Eclogite
- Granulite

A. Metamorphism and Plate Tectonics

Continental crust

Cold sea water descends

Greenschist

Hot water rises

Amphibolite

Greenschist

Hot water rises

Amphibolite

Magma

B. Metamorphism at Divergent Plate Margins

Low P
High T

High P
Low T

High P
Low T

Greenschist

Blueschist

Greenschist
Amphibolite

Greenschist

Granulite

Eclogite

0   50 km

C. Metamorphism at Convergent Plate Margins

FIGURE 5.6
Origin of Metamorphic Rocks

Specimen A

Composition
Muscovite	70%
Garnet	25%

Rock name _____

Origin_____

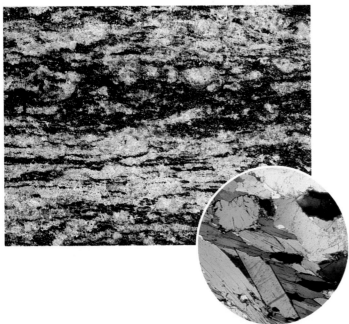

Specimen B

Composition
K-feldspar	30%
Plagioclase	20%
Mica	20%
Amphibole	30%

Rock name _____

Origin_____

Specimen C

Composition
$CaCO_3$	90%
Chlorite	5%

Rock name _____

Origin_____

Specimen D

Composition
Chlorite	80%
Other minerals	20%

Rock name _____

Origin _____

Specimen E

Composition
Quartz	95%
Rock fragments	5%

Rock name _____

Origin _____

Specimen F

Composition
$CaCO_3$	95%
Iron	3%
Chlorite	2%

Rock name _____

Origin _____

GEOLOGIC TIME— RELATIVE DATING

▌ OBJECTIVE

To become familiar with the principles used to determine the relative ages of rock bodies and to use these principles in deciphering the chronological sequence of geologic events.

▌ MAIN CONCEPT

The interpretation of events in Earth's history is based on the *principle of uniformitarianism* (the laws of nature do not change with time).

▌ SUPPORTING IDEAS

1. Rocks are records of events in Earth's history.
2. Relative dating (determining the chronological order of a sequence of events) is achieved by applying the following principles:
   - **a.** superposition
   - **b.** faunal succession
   - **c.** crosscutting relations
   - **d.** inclusions
   - **e.** succession in landscape development

Rock bodies in Earth's crust are not all the same age and were not formed in the same way. They were formed over a vast period of time by a variety of geologic processes and each is a record of a significant event in geologic history. Igneous rocks are records of thermal events, resulting in igneous intrusions or volcanic eruptions. Sedimentary rocks record changing events and environments on Earth's surface, such as the rise and fall of sea level, climatic changes, and changes in life forms. Metamorphic rocks are records of other types of events, such as the collision of tectonic plates and the formation and erosion of mountain belts. Each rock body is much more than just solid material beneath the soil. It is a document of the past, a record of our planet's history.

RELATIVE DATING

Two different concepts of time, and hence two different but complementary methods of dating, are used in geology: relative time and absolute time. *Relative dating* is simply determining the chronological order of a sequence of events. No quantitative or absolute number of days or years is deduced,

only that one event occurred earlier or later than another. *Absolute dating,* in contrast, designates a specific age in units of hours, days, or years. In this exercise, we will be concerned only with relative dating.

Relative dating is important in geology because many physical events—volcanism, canyon cutting, the deposition of sediment, the deformation of Earth's crust—can be recognized in the field. Their relative age can be established by applying several principles of remarkable simplicity and universality. The most significant of these are:

1. *The principle of superposition,* which states that in a sequence of undeformed sedimentary rocks, the oldest beds are on the bottom and the youngest are on the top.

2. *The principle of faunal succession,* which states that groups of fossil plants and animals occur in the geologic record in a definite and determinable order and that a period of geologic time can be recognized by the fossils contained in the rocks.

3. *The principle of crosscutting relations,* which states that igneous intrusions and geologic fractures, such as faults, are younger than the rocks they cut.

4. *The principle of inclusion,* which states that a fragment of a rock incorporated or included in another is older than the host rock.

5. *Succession in landscape development.* Many landforms evolve through a definite series of stages, so that a landscape's relative age can be determined from the degree of erosion. This is especially obvious in volcanic features such as cinder cones and lava flows.

PROBLEMS

This exercise involves analyzing photographs and block diagrams of areas where crosscutting relations, superposition, and stages of landscape development show relative ages of rock bodies that document geologic history. It is an exercise of observation, thought, and interpretation that should reward you with a new awareness of geologic time.

1. Relative Ages of Igneous and Metamorphic Rocks in the Canadian Shield

This low-altitude aerial photograph of part of the Canadian Shield covers approximately three square miles. The area shown is in the subarctic region and was eroded by glaciers during the last ice age. As a result, there is little soil and sparse vegetation, so that bedrock is exposed throughout most of the area. Field mapping indicates that there are three major rock types, each of which appears in distinct shades of gray. The very light gray tones represent granite, medium gray areas are foliated metamorphic rocks, and dark gray areas are basalt dikes.

a. Determine the relative age of these rock bodies.

b. Draw an idealized cross section showing the relations of these rocks.

c. Outline the geologic history of this area. Be sure to include the events involved in the formation of each rock unit, erosion, and evolution of the landscape.

(Department of Energy, Mines, and Resources, Ottawa, Canada)

2. SP Crater, Arizona

a. Draw the boundaries of the major lava flows in the area. Label them according to their relative age (1 = oldest, 2 = next oldest, and so on).

b. How do the flows change with time? Consider surface structures, soil, volcanic cones, and flow margins.

c. How many volcanic cones and cone remnants can you recognize? Label them according to their relative age (1 = oldest, and so on). How do the cones change with time?

d. What is the relationship of flow margins to the development of drainage lines?

(U.S. Department of Agriculture)

3. Geologic Events in the Grand Canyon

The Grand Canyon of the Colorado River is one of the most remarkable exposures of rock bodies in the world. Over a century ago, Major John Wesley Powell made the first boat trip through the canyon and called it "a book of geology." In this exercise, we will attempt to read part of that book.

Four major sequences of rock are exposed in the Grand Canyon. These are shown in the photographs below and on page 71. The canyon is big—nearly 300 mi long. In order to study and interpret its exposed rock bodies, it is necessary to make a composite cross section, such as that shown in (D) on page 71.

a. Carefully study all of the photographs first, to make sure you recognize each major rock body. Then, extend the lines between the major formations across the entire photograph. A cross section and names of formations are shown on page 71.

b. Arrange the following events in their proper chronological order, from oldest to youngest.

(1) Deposition of the Grand Canyon Series

(2) Expansion of the Paleozoic seas and deposition of the Tapeats, Bright Angel, and Muav formations

(3) Erosion of the Grand Canyon

(4) Deposition of the Redwall Limestone

(5) Deposition of the rocks that were later metamorphosed to form the Vishnu Schist

(6) Expansion of the sea over the area and deposition of the Kaibab-Toroweap Limestone

(7) Mountain building and metamorphism to form the Vishnu Schist

(8) Post-Redwall erosion

(9) Tilting and faulting of the Grand Canyon Series

(10) Erosion and exposure of the Vishnu Schist

(11) Intrusion of Zoroaster Granite

(12) Deposition of the Supai and Hermit rocks

(13) Extrusion of basalt and formation of lava dams and lakes in the Grand Canyon

(14) Expansion of a desert and deposition of the Coconino Sandstone

(15) Deposition of the Temple Butte Limestone

(16) Erosion and formation of the Great Unconformity

A. View of the Grand Canyon from Lipan Point, approximately 30 mi east of the Visitor's Center. Three major rock bodies are exposed here. (1) The Vishnu Schist; a complex of metamorphic rocks which form a rugged V-shaped inner gorge (to the left). (2) The Grand Canyon Series consists of 12,000 ft of sandstones, siltstones, and limestones. Basalt (not exposed here) also occurs in this sequence as flows, dikes, and sills. The Grand Canyon Series is tilted 15 degrees to the east. (3) The Paleozoic sequence of rocks consists of nearly 5000 ft of sandstones, limestones, and shales that rest unconformably on the Grand Canyon Series. The rocks are essentially horizontal. Several unconformities occur in the sequence. The formations, from the base to the canyon rim, are: Tapeats Sandstone, Bright Angel Shale, Muav Limestone, Temple Butte Limestone, Redwall Limestone, Supai Sandstone, Hermit Shale, Coconino Sandstone, and the Toroweap-Kaibab limestones.

B. View of the Grand Canyon from Pima Point, near the Visitor's Center. Here, the Grand Canyon Series is missing, and the Paleozoic formations shown in (A) rest unconformably on the metamorphic rocks, called the Vishnu Schist, which forms the V-shaped rugged inner gorge. The Vishnu Schist has been intruded by the pink Zoroaster Granite, which is well exposed several miles east of Pima Point. The Great Unconformity separates the Tapeats Sandstone from the underlying Vishnu Schist.

C. View of the Grand Canyon at Toroweap Valley, approximately 80 miles west of the Visitor's Center. Volcanic activity dominates this scene. A volcano (Vulcan's Throne) is perched on the very rim of the inner gorge, and frozen lava cascades, 3000 ft high, spill into the inner gorge. Remnants of high lava dams, which once blocked the flow of the Colorado River and formed temporary lakes upstream, can be seen clinging to the canyon walls. The Toroweap fault extends up Toroweap Valley and has displaced the Paleozoic rocks by 900 ft.

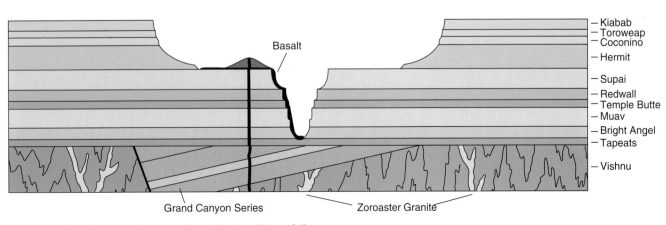

D. Composite Diagram Showing the Relationships of the Major Rock Bodies in the Grand Canyon

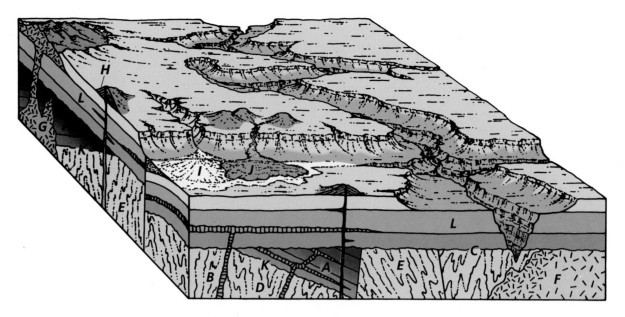

4. Sequence of Events

The diagram above shows several types of rock bodies, unconformities, and crosscutting relationships. Although this is a hypothetical situation, all the relationships between rock bodies and landforms shown in the diagram occur in the Colorado Plateau. The major rock bodies, faults, and unconformities are labeled by letters, each representing a significant geologic event. These are shown in *alphabetical*, not chronological, order in the adjacent column. Study the diagram and list these events in their proper chronological order.

A. Inclined sequence of sedimentary rock

B. Igneous intrusion--basalt

C. Erosional surface

D. Igneous intrusion--basalt

E. Metamorphic

F. Igneous intrusion--granite

G. Igneous intrusion--granite

H. Volcanic cinder cone

I. Alluvial fan

J. Lava flow

K. Erosional surface

L. Horizontal sedimentary sequence

MAPS AND AERIAL PHOTOS

OBJECTIVE

To become familiar with the major types of aerial photographs, remote sensing imagery, and topographic maps and to learn how to analyze and interpret them in the study of Earth's surface features.

MAIN CONCEPT

Topographic maps, aerial photographs, Landsat, and remote sensing imagery are scale models of portions of Earth's surface and are fundamental tools of geologic research.

SUPPORTING IDEAS

1. Aerial photographs are important in geologic work because of the vast amount of surface detail recorded on them.
2. Landsat imagery provides important regional information about Earth's surface features and geologic processes.
3. Various remote sensing devices, such as radar imagery, provide specialized geologic information.
4. Topographic maps show the configuration of Earth's surface by means of contour lines. These maps are of fundamental importance in geology because accurate measurements of both horizontal and vertical distances can be made from them.

The study of Earth's surface features, and the processes that form them, presents a major problem of scale. Mountain ranges, plateaus, and drainage basins are all too large to be seen from any single viewpoint, or even from a thousand different viewpoints. To study such large features, geologists use various types of aerial photography, Landsat images, remote sensing data, and topographic maps. Each is, in a real sense, a type of scale model, showing various aspects of the size, shape, and spatial relationships of Earth's surface features.

The great value of these models is that they provide a regional perspective from a vertical view. They represent the reduction of vast amounts of data to a model, the size of a piece of paper, that can be analyzed and easily managed. Topographic maps, aerial photographs, and remote sensing are, therefore, basic tools for the geologist and are indispensible to the study of landforms and their origin. Just as the microscope gives the biologist a new perspective, or the telescope aids the astronomer, so these models help geologists to study structure and landforms that cannot be seen from viewpoints on the surface.

AERIAL PHOTOGRAPHY AND REMOTE SENSING IMAGERY

Aerial Photographs

Aerial photographs have long been a fundamental tool in geologic studies because they show Earth's surface features in remarkable detail from a vertical perspective. To beginning students, however, a vertical view of Earth's surface is new and unnatural. Many features at first glance may appear simply as unfamiliar or exotic patterns, quite unlike the features we are used to seeing from the ground. Roads, buildings, farmlands, and lakes may be recognizable, but some experience is needed to identify various types of terrain, rock bodies, and other geologic features.

Stereoscopic Viewing. One of the great advantages of aerial photographs is that they can be viewed stereoscopically—in which an image of the landscape appears in three dimensions. This is possible because aerial photographs are taken in sequence, along a flight line with approximately 60% overlap of any two adjacent photos. When two adjacent photos along a

flight line are viewed through a stereoscope, your brain combines the images on the photographs to produce one three-dimensional image of the surface. Hills and valleys appear to stand out in bold relief. Stereoscopic viewing of the landscape is like having a personal view of the landscape from directly above. There is an added advantage because photographs can be used to make notations and for mapping.

Many of the vertical aerial photographs in this manual are printed in pairs called stereograms (Figure 7.1). To view a stereogram, simply place a lens stereoscope over the stereogram so that your nose is directly above the line separating the two adjacent photos (Figure 7.1A). The axis of the stereoscope (an imaginary line connecting the center of each lens) should be perpendicular to the line separating the photos. If stereovision is not attained immediately, rotate the stereoscope slightly, as shown by the arrows in Figure 7.1A, or adjust the lens separation on the stereoscope to fit your eyes better. With a little practice, you can become efficient at stereoscopic viewing, which will provide an important new method for studying Earth's surface features. Figure

A. Using the Stereoscope

7.2 shows some different types of aerial photographs and the common geologic features seen on them.

Landsat and Radar Imagery

On July 23, 1972, the United States launched the first Earth Resources Observation System (EROS) and began collecting repetitive satellite images of Earth's surface. Additional imaging satellites have since been launched in near-polar orbits 570 mi (918 km) above Earth. Landsat images can be manipulated in false color, and further enhanced by computer, to make available a tremendous amount of data about Earth's surface. Other remote sensing images are produced by radar, which is capable of "looking through" cloud cover to record remarkably clear images of surface features. Examples of various types of Landsat and remote sensing images are shown in Figure 7.3.

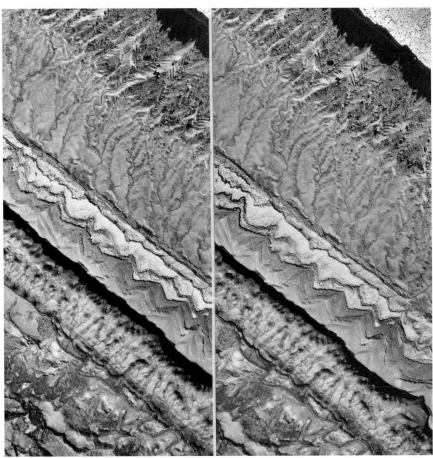

B. Example of a Stereogram. Practice using the stereoscope with this image.
(U.S. Department of Agriculture)

FIGURE 7.1
Viewing a Stereogram

0 | 0.2 miles

A. Low Altitude: sea cliffs, urban subdivision, and interstate highway. Note the detail that is recorded on a low-altitude photograph, including the extent of impermeable surfaces—such as roads, buildings, and parking lots—all of which affect rates of surface runoff and flooding.

0 | 1 mile

B. Intermediate Altitude: city and port facilities. Note the rectangular city-street pattern, circular petroleum storage tanks, ships anchored at dock, railroad tracks, and large warehouse buildings.

0 | 1 mile

C. Intermediate Altitude: volcano and lava flow. Note the cinder cone and summit crater, untouched by erosion. The dark, rough lava flow extends southward from the base of the cone. Older lava flows form elongate ridges nearby.

0 | 1 mile

D. High Altitude: woodland and farms in the Appalachian Mountains of the eastern United States. Note that the rock structure (resistant layers of sedimentary rock) imparts a definite linear texture to the terrain.

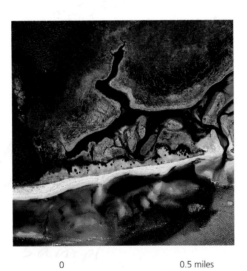

0 | 0.5 miles

E. Low Altitude, Natural Color: shoreline, barrier beach, lagoon, and the open sea. The mainland is covered with dense vegetation, but the tidal flat (gray-orange) that forms the inner shore of the lagoon is mostly a mud flat, cut by tidal channels. A considerable amount of sediment is moving offshore, beyond the beach.

0 | 1 mile

F. High Altitude, Color Infrared: coast, farms, and woodlands. Note the patterns of near-shore currents, indicated by moving sediment (light blue). Various types of vegetation appear red on color infrared photography.

FIGURE 7.2
Examples of Aerial Photography Used in Geologic Studies
(C and F, U.S. Department of Agriculture, Salt Lake City; D and E, courtesy of NOAA, Washington, D.C.)

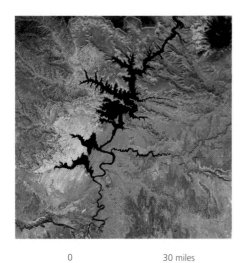

0   30 miles

0   50 miles

0   1 mile

A. Satellite Photograph (Natural Color): Lake Powell area, Utah. Note the dendritic drainage pattern of the tributaries to the Colorado River, accentuated by the contrast of the dark lake water and the red rocks of the plateau country. A large dome, formed by an igneous intrusion, is located in the northeast corner of the image.

B. Landsat Mosaic–Thematic Mapper Image: Chesapeake Bay area. Drainage and coastal features are well expressed as are variations in vegetation types. Ridges of the Appalachian Mountain belt can be seen in the upper left.

C. Landsat Image, Enlarged and Computer Enhanced: mountains, drainage systems, and farmland, Wyoming. This image is enlarged to nearly the scale of an aerial photograph, and many details of the landscape are shown clearly. Note the meandering river and adjacent farmland. Sedimentary rocks inclined toward the south form a series of parallel ridges.

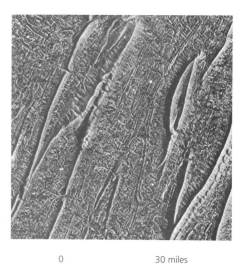

0   30 miles

0   0.5 miles

0   30 miles

D. Landsat Image: San Francisco Bay area, California. This image is false color — vegetation appears red. Note the cities in the bay area, tidal zones, the mud in the bay, and erosion in the coast ranges.

E. Radar Image: Ouachita Mountains. Although a radar image appears to be similar to an aerial photograph, there are significant differences. Note the complete absence of clouds or haze and the lack of detail in the water in the reservoir. Landforms (ridges and valleys) and drainage patterns are very distinct.

F. Radar Mosaic: part of the Cumberland 1:250,000 quadrangle. The advantage of a radar mosaic is that it provides a regional view of the terrain, with excellent detail of the surface features. A large part of a mountain range can be studied from one synoptic view.

FIGURE 7.3
Examples of Landsat and Radar Imagery Used in Geologic Studies
(C and D, Earth Satellite Corporation; B, E, and F, courtesy of EROS Data Center)

TOPOGRAPHIC MAPS

Topographic maps are maps that show the configuration of Earth's surface by means of contour lines. They are constructed to scale, so that distance, directions, areas, elevations, slope angles, and volumes can be measured with great accuracy. These maps are therefore an indispensable tool for geologic studies and are critical to many other endeavors as well (construction, planning, transportation routing, and military operations, for example). Indeed, they are of fundamental importance to the nation's economic development and environmental protection. Topographic maps are accurate terrain models and, as such, show infinitely more than the location of features, distance, and direction. As scale models, they contain a vast store of information that may not be readily apparent to the casual observer. However, the extraction and use of this information are available to anyone with the skill to use this type of map. Users are limited only by their own skills, experience, and knowledge.

Locating Features on a Map

Latitude and Longitude. An important function of a map is to provide the means of defining the location of a point or an area, so a specific position can be found again at a later date, communicated to others, or recorded for scientific, legal, or political purposes. The most effective way to do this is by using a system of grid lines (Figure 7.4). By international agreement, Earth's surface is divided into a series of north–south and east–west grid lines. The north–south lines are called lines of longitude, or meridians. They represent segments of arc on the equator, measured in degrees, minutes, and seconds. The line of zero longitude, the *prime meridian,* passes through Greenwich, England. All other longitude lines are measured as east or west of the prime meridian to the 180-degree line of longitude known as the *international date line.*

The other dimension of the grid is latitude. Lines of latitude circle the globe parallel to the equator, which is zero latitude. Degrees of latitude are measured north or south of the equator up to 90-degrees latitude at either pole. The system of latitude lines also divides the globe into two hemispheres, the Northern Hemisphere and the Southern Hemisphere. Used together, the exact position of any point on Earth's surface can be established.

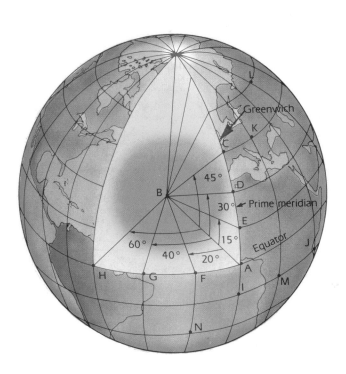

FIGURE 7.4
Measurements of Latitude and Longitude.

The equator forms a great circle on which angles of longitude are measured. The prime meridian is a north–south line through Greenwich, England. Point F, in the Atlantic Ocean, if projected to the center of Earth and back to the prime meridian, would describe angle FBA, which is 20 degrees. All points lying on the north–south line through point F would thus be 20 degrees west of the prime meridian. Point G, near the east coast of South America, if projected to the center of Earth and back to the prime meridian at point A, would describe angle GBA, which is 40 degrees. All points lying on a north–south line through point H would be longitude 60 degrees west. Point J is 40 degrees east longitude.

Angles of latitude are measured on a great circle passing through the poles, with the equator as a reference line. Point C, in Europe, if projected to the center of Earth and back to the equator, would describe angle CBA, which is 45 degrees. All points lying on a line that passes through point C, parallel to the equator, would be 45 degrees north of the equator. This includes the line passing through the northern part of the United States. Similarly, point D, in northern Africa, if projected to Earth's center and back to the equator, would describe angle DBA, which is 30 degrees. All points on a line parallel to the equator passing through point D would be 30 degrees north latitude. Point E is 15 degrees north latitude, and point I is 15 degrees south latitude.

Land Office Grid System. Latitude and longitude lines are excellent tools for large regional reference but are cumbersome when used to subdivide the land surface on a local scale. Most of the United States west of the Ohio and Mississippi rivers has been surveyed and subdivided in accordance with the U.S. Federal Rectangular Survey, or the Land Office Grid System (Figure 7.5). With this system, subdivisions involve establishing an initial point from which all locations of a state or region can be referenced. The latitude line for the initial point is the *base line,* and the longitude line is the *principal meridian.* From the reference point so designated, grid lines are surveyed at 6 mi intervals.

The north–south lines define strips of land known as *ranges* and are numbered (1, 2, 3, and so forth) east and west of the principal meridian (Figure 7.5A). East–west lines establish strips of land, known as *townships,* that are numbered north and south of the baseline. Each square of the grid contains 36 mi^2 and is also referred to as a *township.* Its location is designated by the numbers of the township and by the tiers of ranges it contains. The blue shaded area in Figure 7.5A would be designated Township 3 south, Range 2 west, or simply T3S, R2W.

A township is further subdivided into 36 *sections.* Each section is 1 mi^2 and is numbered as shown in Figure 7.5B. The blue area in Figure 7.5B

would be identified as Section 11, Township 3 south, Range 2 west (Sec 11, T3S, R2W).

The individual sections in turn are divided into quarters and eighths (Figure 7.5C). The large blue area in Figure 7.5C would be identified as the southeast quarter of Section 11, Township 3 south, Range 2 west (SE¼, Sec 11, T3S, R2W). The smaller blue area in Figure 7.5C would be identified as the northeast quarter of the northwest quarter of Section 11, Township 3 south, Range 2 west (NE¼, NW¼, Sec 11, T3S, R2W).

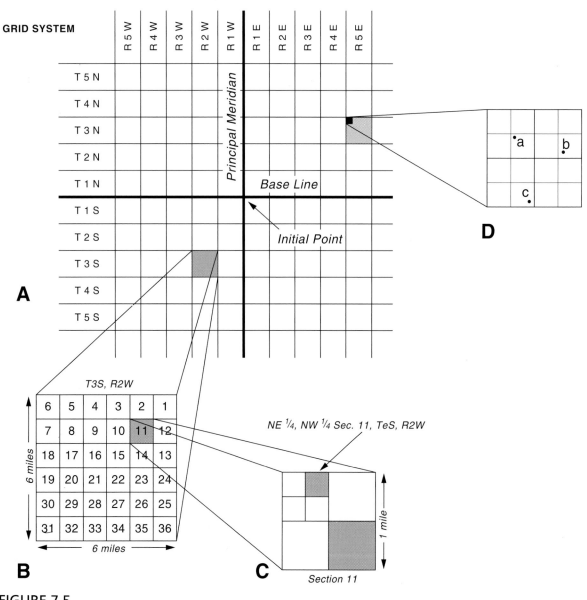

FIGURE 7.5
The Land Office Grid System

Scale

Because a topographic map is intended to be an accurate scale model of a portion of Earth's surface, the features shown on it will differ from those on Earth in size but not in shape. A map's scale is the ratio between the size of an area represented on the map and the size of the actual area on the surface. Three expressions of scale are commonly used with topographic maps: (1) fractional, (2) graphic, and (3) verbal.

1. The fractional scale is expressed as a simple fraction (1/62,500) or as a proportion (1:62,500). These expressions mean that one unit of distance on the map is equal to 62,500 of the same unit on the ground.

2. A graphic scale is a line or bar divided into a number of segments that represent units of length (feet, kilometers, or miles) on the ground. In a sense, this scale is like a ruler that can be used for measuring distance.

3. A verbal scale is a simply worded statement of the relationship between map distance and ground distance (1 inch equals 1 mile).

The U.S. Geological Survey publishes maps on a variety of scales to meet the needs of supplying both detail and broad coverage. (See Figures 7.6 and 7.7). The survey system of subdividing areas for map coverage uses the universal coordinate lines of latitude and longitude. Maps bounded by latitude and longitude are known as *quadrangles*. The system provides quadrangles of different sizes and various scales, like those shown in Figure 7.7.

Directions

Unless otherwise indicated, true north is toward the top of the map. Meridians and lines of latitude serve as convenient reference lines, as they run north–south and east–west, respectively. In establishing the direction from one point to another, it is convenient to do so with reference to north and south. This is done in a general way, such as saying that town X is 6 mi northwest of town Y. In more precise work, the angle east or west of a north–south line is measured with a protractor. For example, on the Kaaterskill map (page 101), Parker Mountain is N 52° W of the

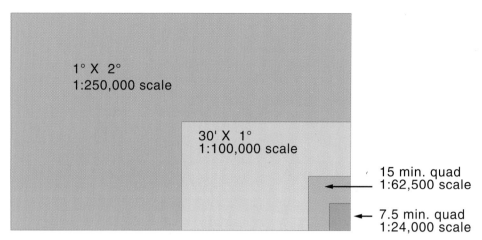

FIGURE 7.6
Relative Size of Standard U.S. Geological Survey Topographic Quadrangles

town of Haines Falls, and Roundtop Mountain is N 80° E of Elka Park. True north and magnetic north are both illustrated near the bottom of a standard quadrangle map, and their difference, or declination, is indicated. When using a topographic map in the field, the magnetic compass direction must be corrected for declination to obtain true north.

Determining Elevations

Elevation refers to height (in feet or meters) above sea level and is essentially synonymous with *altitude*. Specific elevations, established by highly accurate surveys, are referred to as *benchmarks* and are shown on topographic maps in various ways. Typical benchmark locations are at the center of a town, on a hilltop, and at the bottom of a depression. The benchmarks are printed in black on a map. Spot elevations, which are less accurate, are printed in brown.

In addition, the approximate elevation of any point on the map can be determined from the contour lines. Each contour line connects all points of the same elevation. For points between lines, elevations can be interpolated. For example, a point midway between the contours of 1240 ft and 1260 ft would probably have an altitude of 1250 ft, and a point located just below the 1260-ft contour line would probably be at an elevation close to 1258 or 1259 ft. Such approximations are based on the assumptions (1) that the slopes

have a constant gradient and (2) that the elevation is proportional to the horizontal distance. These assumptions are not always true, but a careful study of slope trends usually permits an accurate estimation of the elevations between contours.

Relief is the difference in elevation between high and low points. You can easily determine the local relief of an area by subtracting the lowest elevation from the highest elevation.

Height and *depth* are measurements made relative to some local feature. For example, a monument might be 555 ft high, relative to the ground, but have an overall elevation (at its top) of 1555 ft, if the ground is 1000 ft above sea level.

Colors and Symbols

If maps showed all of the natural features, as well as the marks of civilization within any given area, they would be so cluttered as to be useless. For the sake of clarity, symbols are used to indicate a variety of cultural features (e.g., buildings, political boundaries, roads, and railroads). Contour lines are printed in brown. Streams, rivers, lakes, and other bodies of water are printed in blue. On some maps, features of special importance are shown in red, and vegetation is in green. Map symbols used by the U.S. Geological Survey appear inside the back cover of this book.

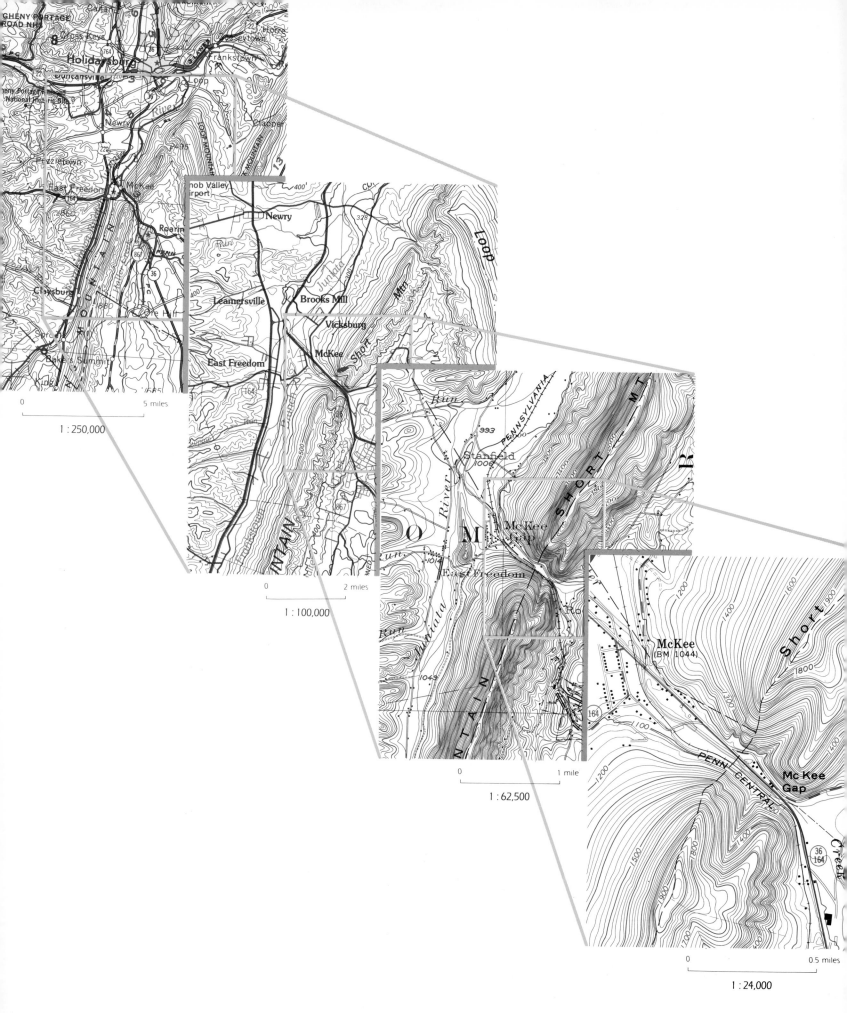

FIGURE 7.7
Examples of Scales of Topographic Maps
(U.S. Geological Survey)

Contour Lines

The great value of topographic maps is that they show the shape and elevation of the land surface. This is done by means of contour lines. A contour line is an imaginary line on Earth's surface connecting points of equal elevation. A contour line can also be described as a line traced by the intersection of a level surface with the ground. Natural expressions of contour lines are elevated shorelines, cultivated terraces, and patterns produced by contour plowing.

To help understand the idea of contours, consider an island in a lake and the patterns made on it when the water level recedes (Figure 7.8). The shoreline represents the same elevation all around the island and is thus a contour line (Figure 7.8A). Suppose that the water level of the lake drops 10 ft and that the position of the former shoreline is marked by a gravel beach (Figure 7.8B). Now there are two contour lines—the new lake level and the old stranded beach—each depicting accurately the shape of the island at these two elevations. If the water level should continue to drop in increments of 10 ft, with each shoreline being marked by a beach, additional contour lines would be formed (Figures 7.8C and 7.8D). A map of the raised beaches is, in essence, a contour map (Figure 7.8E).

An aerial photograph of an island in the Great Salt Lake, Utah, is an excellent example of the events just described. (Figure 7.9A). Former shorelines of high lake levels are clearly seen around the island as natural contours. A sketch of the shorelines is a contour map (Figure 7.9B).

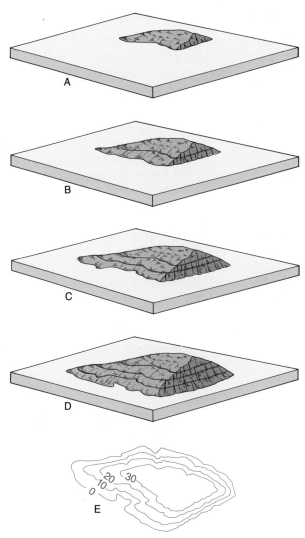

FIGURE 7.8
Elevated Shorelines as Contour Lines

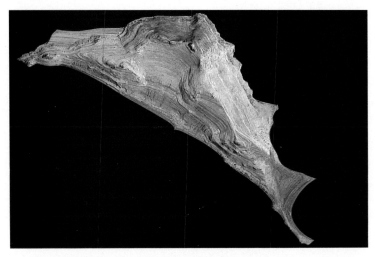

A. Aerial Photograph of Fremont Island, Great Salt Lake, Utah

FIGURE 7.9
Contour Lines Formed by Elevated Shorelines
(A, U.S. Department of Agriculture, Salt Lake City)

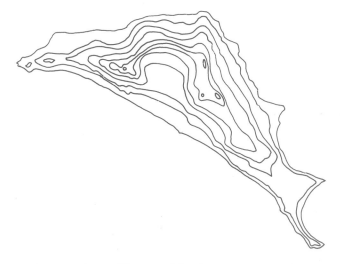

B. Contour Lines of Fremont Island

Characteristics of Contours

Learning to read contour maps effectively is not an easy task. Nature has few lines reminiscent of contour lines, and the visualization of a three-dimensional surface represented only by contours requires careful study and practice. After some training, however, you should be able to visualize the landscape represented by a contour map. You will obtain more quantitative information from a contour map than from any other type of map or image.

The following basic rules will help you get started:

A. The difference in elevation between adjacent contours is constant on any given map and is referred to as the *contour interval,* or CI. The most frequently used contour interval on 7½-min and 15-min quadrangles is 20 ft.

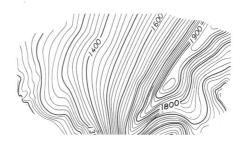

B. Contour lines trend up a valley, cross the stream, and extend down the valley on the opposite side. The lines thus form a V pattern, with the apex of the V pointing upstream.

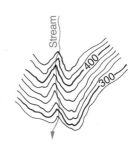

C. Contours never cross or divide. They may appear to merge to express a vertical cliff, but in reality they are stacked one on top of another and only appear to touch. The spacing of contour lines reflects the gradient, or slope. Closely spaced contours represent steep slopes. Contours spaced far apart represent gentle slopes.

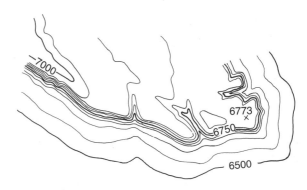

D. Hills and knobs are shown as closed contours.

E. Closed depressions (basins without outlets) are shown by closed contours with hachures (short lines) pointing downslope.

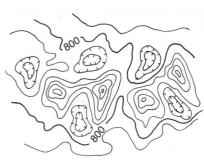

FIGURE 7.10
Characteristics of Contour Lines

Drumlins   Sink Holes   Sand Dunes   Volcanos

Dissected Plateau   Rolling Hills   Fold

FIGURE 7.11
Geologic Features Expressed by Contours.
Many geologic features have specific shapes that are well expressed by contour patterns. Some of the more obvious are shown above.

DIGITAL SHADED RELIEF MAPS

How to portray the surface of the land realistically is a problem mapmakers have struggled with for centuries: how to render a three-dimensional landscape on a two-dimensional map. All traditional solutions, including contouring, which is partly implemented by machine, have been artistic. Among the various techniques cartographers have developed for rendering landforms on maps are hachuring, elevation tinting, pictoral relief, and shaded relief. However, the details of the landscape are much too complex to be recorded accurately and economically over large areas using any of these means.

The problem was recently solved by the computer. Topography need no longer be mapped symbolically using hand-drawn schematic topographic features, or subjectively using artistic shading. Now the computer can represent landforms as they actually are and portray terrain in the infinite varieties of form that constitute the true landscape. The resulting image is a digital *physiographic map* in vertical perspective. Although shaded relief maps may look deceptively like a satellite image, an aerial photograph, or a radar image, they are not created directly from any of these sources. Digital shaded relief maps are made from millions of elevation points derived by digitizing elevation contour lines from conventional topographic maps. Theoretical brightness values (reflected light intensity) are based on the mathematical relationships between sun position and local ground slope and slope direction. The resulting shaded relief map may be printed in black and white, or color tints may be assigned to the various elevation intervals to create a color shaded relief map.

Computer-generated maps such as these offer several advantages for the visual study of the landscape. Above all, these images portray landforms accurately, disclosing their true complexity.

Features can be seen not only in great detail, but also in their broad, regional context. In addition, digital shaded relief maps lack the distortion inherent in aerial photographs and radar images and are free from the vegetation, clouds, and cultural features that mask topography on satellite images. In addition, stereo pairs in shaded relief can be created digitally and sun position can be varied to obtain different aspects of the same area.

Digital maps can be assembled on various scales, ranging from local areas to the entire globe. Elevation data from the sea floor can be integrated with elevations on the continents to produce a map of the planet's entire solid surface. An example of a digital shaded relief map is shown on pages 198 and 199.

FIGURE 7.12
Shuttle Radar Map of Area in Argentina
(Jet Propulsion Laboratory)

SHUTTLE RADAR MAPS

The Shuttle Radar Topographic Mission was launched on February 11, 2000, and, during a period of 11 days, it collected enough data to generate the most accurate and complete topographic map of Earth's land surface ever produced. The results are digital topographic maps similar to the digital shaded relief maps derived from digitizing topographic maps previously described. The difference is in how the elevation data were collected and areas covered and the details shown. The collected radar data covers nearly 80 percent of Earth's land surface, home to nearly 95 percent of the world's population.

The imaging radar works like a flash camera in that it provides its own energy to illuminate an object and take a picture. The flash camera sends out a pulse of light waves (the flash) and records the light reflected back to it on film. Instead of light waves, imaging radar sends out radio waves which are reflected back to an antenna and recorded on digital computer tapes which can be processed into an image. The key advantage of radar is that it can "see" through clouds and darkness. Imaging radar was used by NASA's *Magellan* spacecraft to produce the spectacular pictures of the surface of Venus. An example of a Shuttle Topographic image is shown above.

This image is of part of Patagonia, Argentina, and shows a spectacular landscape formed by volcanoes, river systems, lakes, wind, and mass movement. Interesting features include basalt-capped mesas with sinkholes (lower center), arcuate beach ridges formed by an ancient lake (upper center), young volcanic cones (right), and inverted topography (right center).

Colors show elevations: blue (lowest) to white (highest). Total relief in this area is 3600 ft.

SPECIAL MAPS

A wide variety of special maps are made by the U.S. Geological Survey and by various private institutions such as the mining and petroleum industries, oceanographic institutions, land use planning, and environmental protection agencies. Examples include photoimage maps and geophysical maps showing measurements of magnetism, gravity, and radioactivity.

Of particular interest to geologists are new maps of the sea floor derived from depth soundings and satellite measurements of the sea surface. Some examples of these special maps are shown in Figure 7.13.

A. Orthophoto Maps. Distortion-free multicolored photographic images that have no contours and minimal labels.

B. Geophysical Maps. Multicolored maps that show results of surveys to measure magnetism, gravity, or radioactivity, etc. Contour lines on this map show equal intensity of magnetic forces.

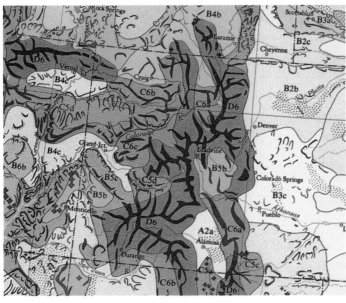

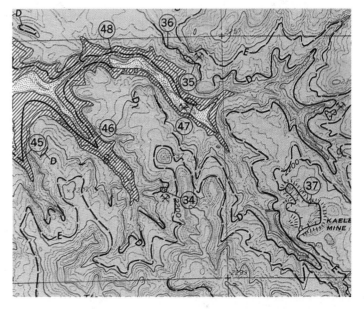

C. Thematic Maps. Maps that show specific features, such as tectonic features, engineering aspects, soils, etc. This particular map shows "classes of land surface forms."

D. Coal Investigation Maps. Multicolored maps showing coal resources. Similar maps may be made showing locations of other resources, such as oil, gas, sand and gravel, etc.

FIGURE 7.13
Examples of Special Maps Used by Geologists

FIGURE 7.14
Computer-Generated Model of Mount St. Elias, Alaska Area
Note the glacial peaks, glaciers, moraines (lower right), and the flow marks in the large piedmont glacier (lower left).
(U.S. Geological Survey)

Computer-Generated Terrain Models

We can now store vast amounts of data contained on topographic maps and remote sensing images in a computer and retrieve the data in various forms. One useful form is to have the computer construct a terrain model in perspective (Figure 7.14). The vertical scale can be exaggerated and the model rotated so the observer can view the terrain from any perspective. This permits the study of many subtle surface features that might not be readily apparent from maps alone.

PROBLEMS

1. Study the examples in Figures 7.2 and 7.3, and list the advantages and limitations of the following types of remote sensing imagery:

    a. Low-altitude aerial photos

    b. Color infrared aerial photos

    c. Black and white Landsat imagery

    d. Color computer-enhanced Landsat imagery

    e. Radar imagery

2. Using a lens stereoscope, study the stereograms on page 88. What are the advantages of stereo photos in geologic studies?

3. Study Figure 7.4 and determine the latitude and longitude of points K, L, M, and N.

4. Identify the locations of points a, b, and c in Figure 7.5D.

5. Viewing Stereoscopic Contour Maps

Perhaps the best way to learn to read a contour map is to study contour lines in stereoscopic vision (see the opposite page). This should help you to visualize more easily surface features such as hills, valleys, cliffs, gentle slopes, and depressions when they are portrayed on conventional nonstereo maps. Stereographic contours give you the opportunity to visualize the three-dimensional model that is the focus of regular topographic maps.

An effective method is first to view the maps with a stereoscope and then to close one eye, so that you see two dimensions only. Study a specific feature such as a hill, cliff, or gentle slope.

With sufficient practice, your experience with three-dimensional contour maps should permit you to visualize easily the three-dimensional aspect of a traditional contour map.

Four stereo-topographic maps are shown on the opposite page, together with a list of landforms that appear on each. Using appropriate numbers, label the landforms on the stereo-maps.

a. Hills and Valleys

(1) Hills

(2) V-shaped contour pointing upstream

(3) Highest point on the map

(4) Steepest slope

b. Mesas and Buttes

(1) Top of the mesa

(2) Steep cliff

(3) Lowest point on the map

(4) Highest surface on the map

(5) Valleys cut in the surface of the mesa and butte

c. Canyons

(1) Undissected plateau

(2) Head of the canyon

(3) Tributary canyon

(4) Steepest canyon wall

(5) Minor valleys cut into the plateau

d. Closed Depressions

(1) Lowest depression

(2) Highest point

(3) Hills

(4) Show closed depressions with hachure lines

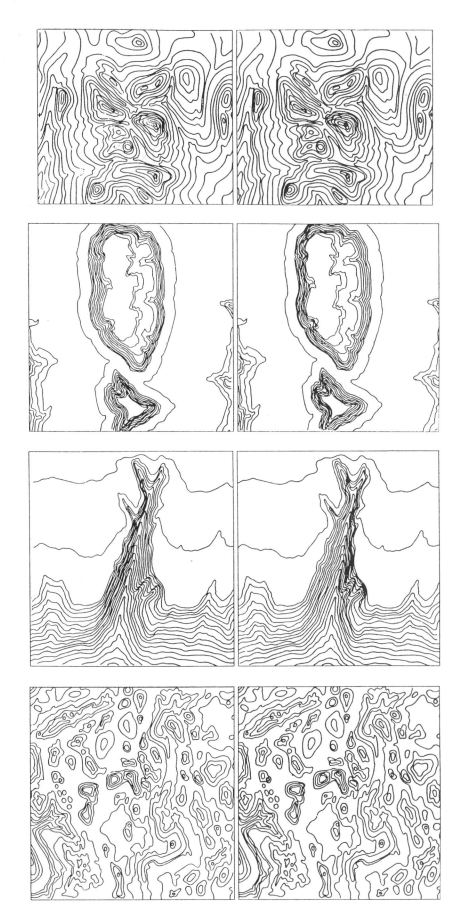

(Courtesy of Gubbard Scientific)

87

6. Comparing a Topographic Map With Stereoscopic Photographs

The Menan Buttes area in southern Idaho (A) offers an excellent opportunity to study the way in which simple topographic forms are expressed by contour lines. By carefully examining with a stereoscope, aerial photographs of the buttes (B) and by comparing the stereoscopic image with the topographic contours, you can further develop your ability to visualize the landforms expressed on a topographic map. The following problems will help you compare the map with the photograph:

a. Study the volcano of the Menan Buttes. The slopes of the crater differ significantly from one side to the other. How are steep slopes expressed by contour lines?

b. How are the more gentle slopes beyond the base of the volcano expressed by contour lines?

c. How are the rugged slopes in the upper left part of the interior of the crater expressed by contour lines?

d. How is the closed depression of the crater expressed by contour lines?

e. Study the floodplain area. How is a relatively flat surface expressed by contour lines?

f. Sketch contour lines on the photo shown in the stereoscopic view. Work directly on the photograph and compare your results with the topographic map.

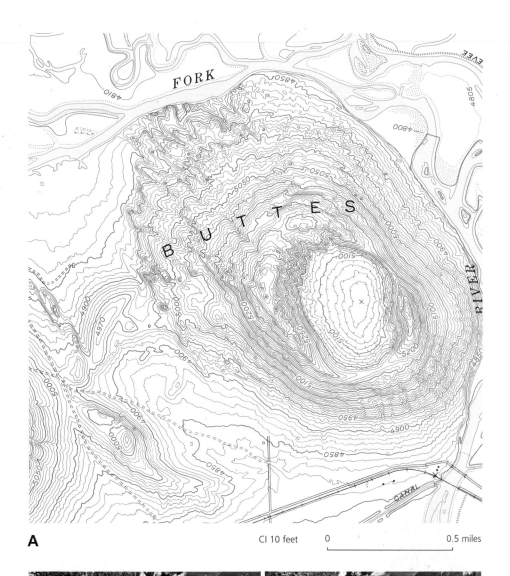

A

CI 10 feet   0   0.5 miles

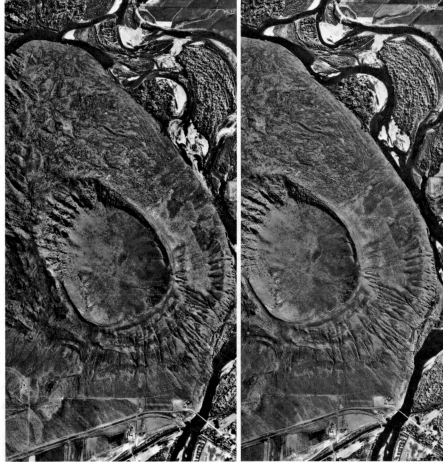

B

(U.S. Department of Agriculture)

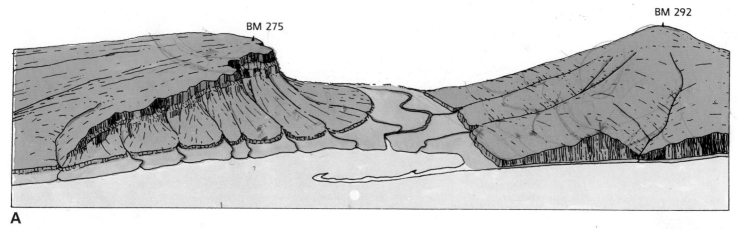

A

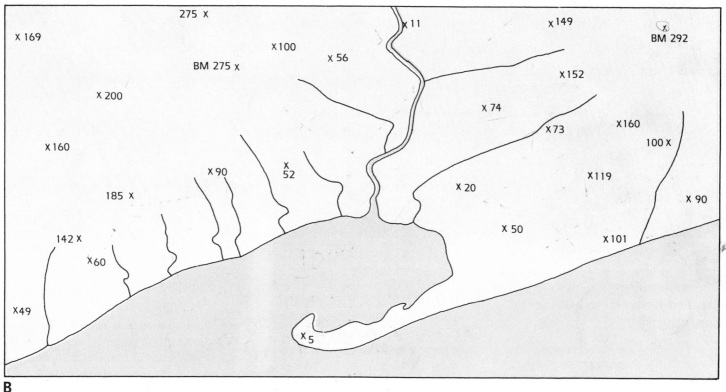

B

(U.S. Geological Survey)

7. Constructing a Contour Map from Established Elevations

Contour maps were originally constructed by surveying the location and elevation of a number of points in the field. The surveyor then sketched contour lines in the field by extrapolating between the surveyed points. Today, maps are created and updated from stereoscopic aerial photographs, and much information is stored, manipulated, and retrieved by computer. This exercise of constructing your own topographic map from surveyed points is extremely useful; it will help you to observe and understand the surface features depicted on a topographic map.

a. Construct a contour map of the landforms shown above. Use a contour interval of 20 ft.

Suggested procedure: First, take time to study the various landforms shown in the sketch (A). Note that the river flows through a low valley and into a bay partly enclosed by a sand spit only 5 ft above sea level. Both sides of the valley are terraced. The hill on the right has been eroded to form a gradual slope above a wave-cut cliff. The hill on the left is an inclined tableland and is crossed by a few shallow valleys. A steep cliff formed on the left hill faces the river and the sea. All of these features should be shown by the contours on your map. As you begin, it may be advantageous to sketch lightly, with a soft pencil, a number of contour lines in perspective directly on the sketch (A). In placing contours on the map, start near the edge, at the lowest elevations, and work up the major streams. Pay particular attention to the established elevation points, and be sure that all contour lines are in harmony with them. To find the appropriate position of a contour line between two control points, study the picture of the slope and estimate the map position of that slope section.

(U.S. Department of Agriculture, Salt Lake City)

0   0.5 miles

8. Constructing a Contour Map From Stereoscopic Aerial Photographs

a. Construct a contour map of the island area shown in stereoscopic view. Use a contour interval of 200 ft.

Suggested procedure: Note the configuration of the shoreline, and remember that the water level is a horizontal surface and that all contours must be parallel to it. Note the elevations established by a survey (printed in white). Use a light lead pencil because you may need to do some erasing before you are satisfied with the result. It is generally best to start with lower elevations and to work up the major streams. Label every fifth contour line.

State briefly how you could increase the amount of topographic detail shown on your map.

9. Constructing a Topographic Profile

a. Construct a topographic profile along line A–A′ across the map in (B) on the opposite page. Topo-graphic maps present a view of the landscape as seen from directly above, an excellent perspective from which to examine regional relationships. This view is, however, unnatural to us, for we are accustomed to seeing hills and valleys from a horizontal perspective. In detailed studies of landforms, it may be desirable to construct a profile, or cross section, through certain critical areas so that various features can be analyzed from a more natural viewpoint. Such a profile can be constructed quickly and accurately along any straight line on a contour map, as illustrated in (A) on the opposite page. The procedure is as follows:

- Lay a strip of paper along the line for which the profile is to be constructed.

- Mark on the paper the exact place where each contour, stream, and hilltop crosses the profile line.

- Label each mark with the elevation of the contour it represents. If contour lines are closely spaced, it is sufficient to label only the index contours.

- Prepare a vertical scale (C) on profile or graph paper by labeling horizontal lines to correspond to the elevation of each index contour line.

- Place the paper with the labeled marks at the bottom of the profile paper and project each contour onto the horizontal line of the same elevation.

- Connect all of the points with a smooth line.

Obviously, the appearance of the profile will vary, depending on the spacing of the horizontal lines on the profile paper. If the vertical scale is the same as the horizontal scale, the profile will be nearly flat. For this reason, the vertical scale is usually exaggerated to show local relief.

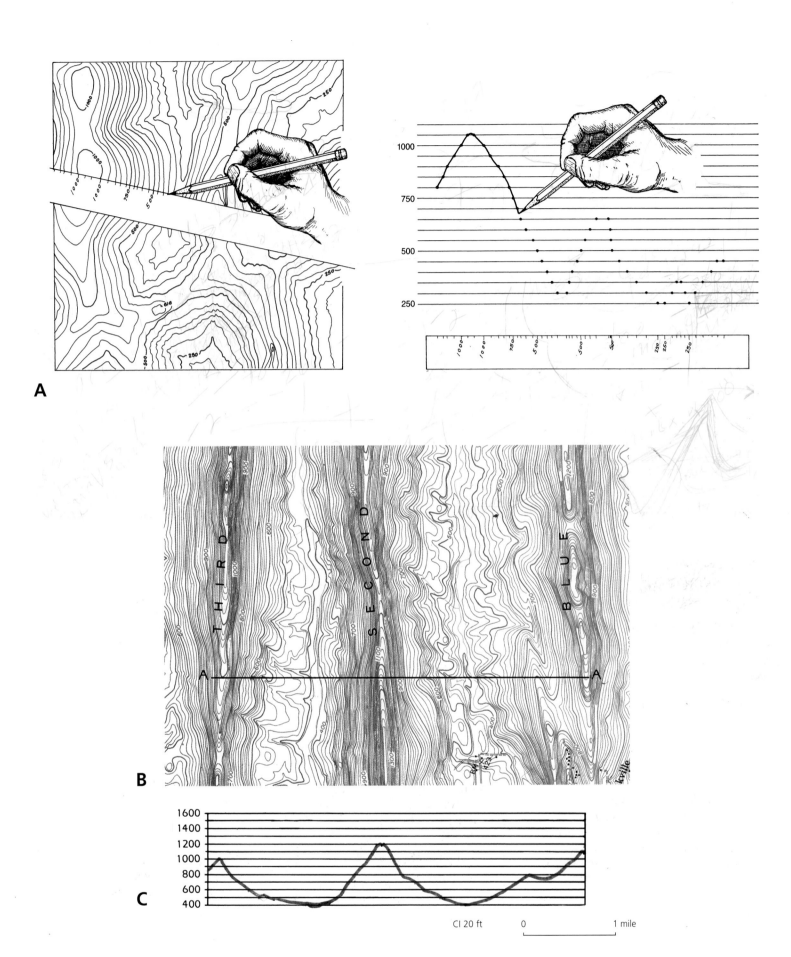

A

B

C

CI 20 ft   0   1 mile

0   100   200   300   400   500 miles

(U.S. Geological Survey)

FIGURE 8.1
Shaded Relief Map of the United States
(U.S. Geological Survey)

94

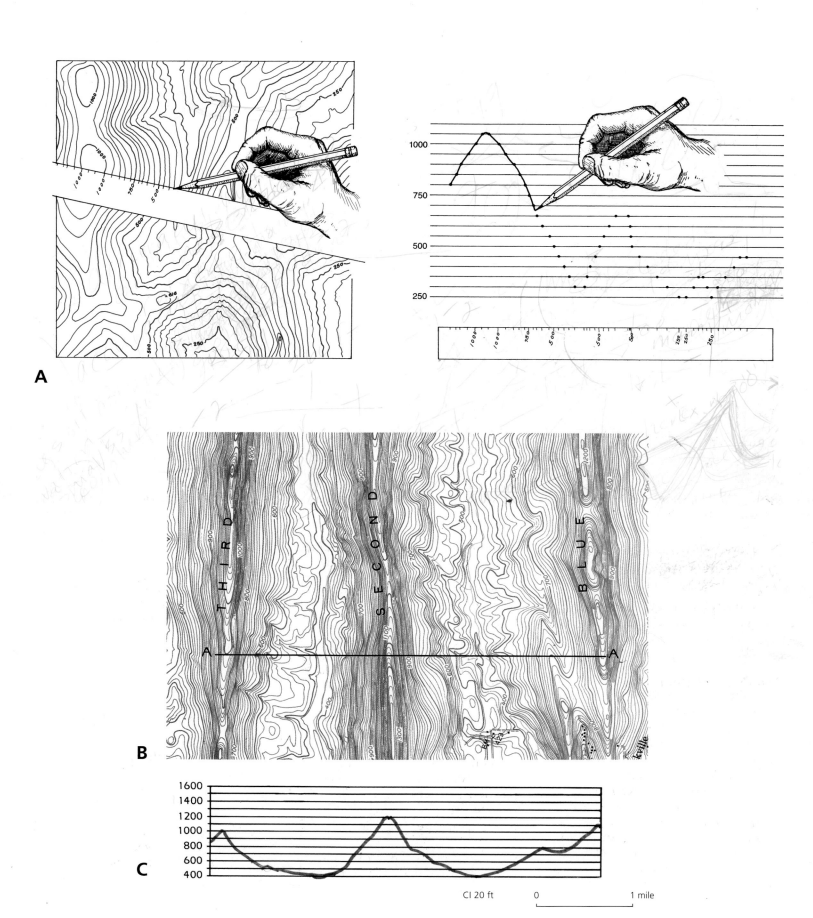

A

B

C

CI 20 ft   0   1 mile

LANDFORMS OF THE UNITED STATES

■ OBJECTIVE

To review the relative size and spatial relationships of the major physiographic provinces of the United States in order to understand better the regional setting and geologic significance of local landforms.

■ MAIN CONCEPT

The major provinces of the United States can be recognized on a landform map on the basis of elevation, topography, patterns of surface features, and origin.

■ SUPPORTING IDEAS

Landforms, river systems, and coasts commonly have many distinctive characteristics that reflect certain facts about their origin and history.

PHYSIOGRAPHIC PROVINCES

The major physiographic provinces of the United States can be classified on the basis of landforms, structure, rock types, and geologic history. A summary of the major divisions follows. As you read this material, locate the various provinces on the relief map on pages 94–95 and study their landforms and boundaries.

Interior Lowlands

The Interior Lowlands extend from Ohio to Kansas. The surface of this region is generally 500–1000 ft above sea level. This province is part of the stable platform in which the rocks are mostly Paleozoic sedimentary strata, warped into broad domes and basins. In several areas, Precambrian crystalline rocks are exposed in the core of eroded domes.

Great Plains

The Great Plains extend from Kansas and the Dakotas westward to the front of the Rocky Mountains. Most of the area is 1000–2000 ft above sea level. This province also is part of the stable platform and is covered with horizontal Mesozoic and Cenozoic sedimentary rocks.

Appalachian Mountains

The Appalachian Mountains are an old mountain belt extending from central Alabama to eastern Canada. The belt is subdivided into four major provinces: (1) the Appalachian Plateau, (2) the Valley and Ridge, (3) the Blue Ridge, and (4) the Piedmont.

1. *The Appalachian Plateau* is a plateau located west of the Valley and Ridge province and is dissected mostly by the drainage system of the Ohio River. The area is underlain by gently dipping sedimentary rocks and is more than 1000 ft above sea level.

2. *The Valley and Ridge* is an elongate belt of folded Paleozoic sedimentary rocks that are deeply eroded. The resistant rock layers have formed extensive zigzag ridges, and the nonresistant formations have been eroded to form intervening valleys.

3. *The Blue Ridge* is an elongate mountainous area located east of the Valley and Ridge and extending from northern Georgia to southern Pennsylvania. The rocks are mostly complex metamorphics.

4. *The Piedmont* is an eroded, rolling upland carved on highly deformed igneous and metamorphic rocks. It lies between the Coastal Plains on the east and the Blue Ridge on the west.

Coastal Plains

The Coastal Plains are a wide lowland extending from New York to Mexico. The area is generally less than 500 ft above sea level and consists of gently dipping sedimentary rocks eroded to form a series of cuestas and strike valleys.

New England

The New England province is a mountainous area eroded on metamorphic and igneous rocks. The region has been glaciated.

Rocky Mountains (Cordilleran)

The Rocky Mountains are part of a complex mountain range located west of the Great Plains. The range is subdivided into a number of provinces based on structure, rock type, and the topography of the area.

1. *The Northern Rockies* are folded and faulted mountains, composed of sedimentary, metamorphic, and igneous rocks. They extend from central Idaho to Canada.

2. *The Southern Rockies* consist of north–south trending, elongate crustal upwarps that extend from central New Mexico to Wyoming. Precambrian crystalline rocks are exposed in the core of the upwarps.

3. *The Central Rockies* are faulted and upwarped mountain ranges in central Utah and western Wyoming. Precambrian rocks are commonly exposed in the core of the upwarps.

Basin and Range

The Basin and Range province consists of numerous ranges that are mostly the result of block faulting, separated by basins filled with sedimentary deposits of Cenozoic age. This province extends throughout Nevada, western Utah, and the southern parts of California, Arizona, New Mexico, and west Texas.

Colorado Plateau

The Colorado Plateau consists of nearly horizontal or gently tilted layers of sedimentary rock. It is located in Utah, Arizona, New Mexico, and Colorado, and is dissected into large canyons by the drainage system of the Colorado River. Volcanic rocks are common along the plateau's margins.

Columbia Plateau

The Columbia Plateau, located in eastern Washington and Oregon and southern Idaho, is composed of extensive basaltic lava flows.

Cascade Mountains

The Cascade Mountains, a north–south trending chain of recent composite volcanoes, extends from northern California to the Canadian border.

Sierra Nevada Mountains

The Sierra Nevada Mountains, a tilted fault block of granitic batholiths, is in eastern California.

Pacific Coastal Ranges

The Pacific Coastal Ranges are complexly folded and faulted masses of sedimentary and metamorphic rocks. The ranges are located in the Pacific coastal states.

Digital Shaded Relief Map of the United States

The digital shaded relief map in Figure 8.1 was published by the U.S. Geological Survey in 1999. The great value of this map is that it shows landforms and tectonic phenomena in vivid detail. These features are so numerous that only a small sample can be discussed here. All of the general surface characteristics of the major physiographic provinces of the United States are crisply rendered in shaded relief. Some of the most obvious are (1) the distinctive structural control of the Appalachian Mountains, (2) the fault blocks of the Basin and Range, (3) the west tilt of the Sierra Nevada Mountains, and (4) the flat coastal plains. The complexity of the Rocky Mountain system and the Pacific Border province is also portrayed in remarkable detail. The contrasting structure and geomorphic styles between the passive continental margin of the eastern United States and the active plate margin of the west, where the North American and Pacific crustal plates are colliding, is a most striking regional feature. The differences are clearly reflected in the coastlines: the Atlantic and Gulf coasts developed on gently sloping continental shelf, and the steep rugged coast of the Pacific formed in a tectonically active zone.

Only slightly less obvious are various smaller features that, if familiar to the viewer, are easily discernible. The broad expanse of the Mississippi River floodplain and the flat surface of the Llano Estacado (Staked Plains) of western Texas are among our country's notable flat surfaces. The Coteau des Prairies in the Dakotas, a flatiron-shaped plateau some 200 mi long and pointing northward toward Canada, is perhaps the most striking feature in the midcontinent. The smooth, ice-scoured lowlands that flank the Coteau des Prairies were occupied by lobes of the last ice sheet. Elsewhere, even smaller features stand out, expressing their own unique style of structure and topography. Sutters Butte volcano, in the northern part of the Great Valley of California, and many of the low volcanic shields of the Snake River Plains are easily seen.

Study the major landforms in the area in which you live and marvel at how accurately they are displayed. A major strength of this map is not so much its expression of the obvious, but rather the detail with which it depicts features that are subtler or less familiar.

PROBLEMS

1. Study the map in Figure 8.1 and locate the following features:

a. Wisconsin Highlands: A gentle domal upwarp with crystalline rocks exposed in the core. Its elevations range between 1000 and 2000 ft.

b. Adirondack Mountains: A large, circular domal upwarp in northern New York in which complex Precambrian igneous and metamorphic rocks are exposed.

c. Black Hills: A large domal upwarp near the Wyoming–South Dakota border in which Precambrian igneous and metamorphic rocks are exposed.

d. Bighorn Mountains: A domal upwarp in north-central Wyoming.

e. San Andreas Fault, California

f. Grand Canyon, Arizona

g. Snake River Plains, Idaho

h. Death Valley, California

i. Mississippi Delta

j. Great Valley of California

2. Briefly compare and contrast the eastern and western coasts of the United States.

3. How is the Idaho Batholith, located north of the Snake River Plains, expressed topographically?

4. What evidence suggests that glacial lakes once extended southwest of Lake Erie?

5. What evidence suggests that a lobe of the glacier once advanced southwest of Saginaw Bay, Michigan?

0   100   200   300   400   500 miles

(U.S. Geological Survey)

FIGURE 8.1
Shaded Relief Map of the United States
(U.S. Geological Survey)

94

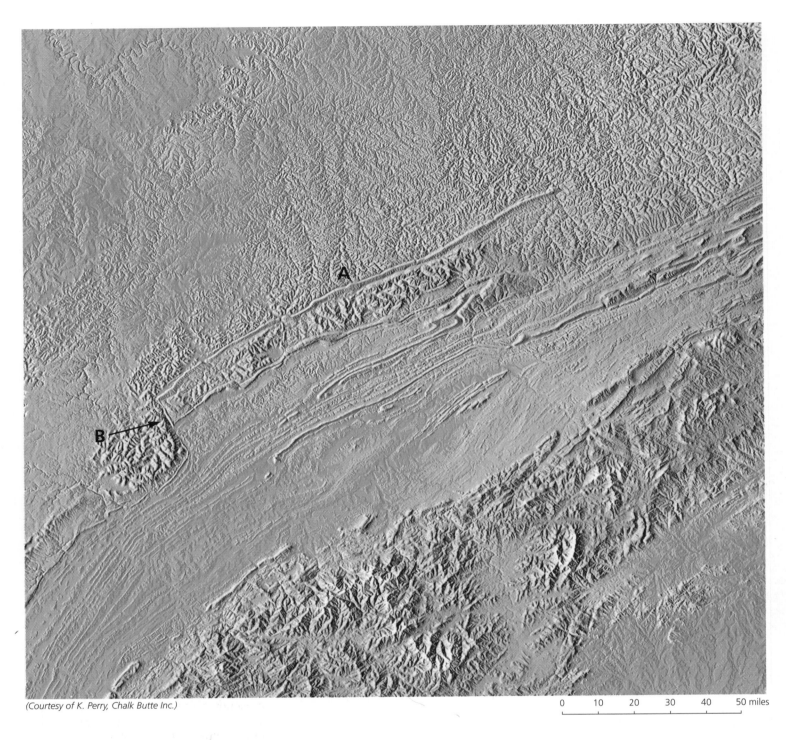

(Courtesy of K. Perry, Chalk Butte Inc.)

0   10   20   30   40   50 miles

6. Digital Shaded Relief Map of the Southern Appalachian Mountains Area

Locate this area on the map in Figure 8.1 and study the boundaries of the major physiographic provinces. Parts of the Coastal Plains, Piedmont, Blue Ridge, Valley and Ridge, Cumberland Plateau, and the Interior Lowland are shown on this map. Draw the boundaries between these provinces. Are the boundaries sharp and well defined or are they gradational?

a. How does the topography of the southern Valley and Ridge province differ from that of the northern Valley and Ridge? Refer to the map on pages 94–95.

b. Study the drainage system (network of valleys) in the western part of this area. What suggests that the bedrock there is nearly horizontal?

c. What produced the long, deep, narrow, straight ridge (point A)?

d. What produced the linear feature separating different terrains at point B?

e. Study the drainage patterns in the eastern part of the area (various shades of green). What is the structure of the bedrock on the Coastal Plains?

STREAM EROSION AND DEPOSITION

▪ OBJECTIVE

To understand the processes of stream erosion and deposition, and to recognize the landforms these processes produce.

▪ MAIN CONCEPT

Rivers erode the landscape by (1) downcutting, (2) headward erosion, and (3) slope retreat. Deposition of sediment transported by a river system produces (1) meanders and point bars, (2) natural levees, (3) backswamps, (4) stream terraces, (5) deltas, and (6) alluvial fans.

▪ SUPPORTING IDEAS

1. The downcutting of stream channels results from the abrasive action of sand and gravel.
2. Headward erosion occurs at the head of all tributaries and causes the drainage network to migrate upslope.
3. Slope retreat is accomplished by a variety of types of mass movement and by the headward erosion of minor tributaries.
4. A river flowing over a low gradient deposits part of its load to form natural levees, point bars, and backswamps.
5. When a river enters a lake or an ocean, it deposits most of its sediment load to form a delta. When a river flows into a dry basin, it deposits most of its sediment load to form an alluvial fan.

PROCESSES OF STREAM EROSION

Running water is by far the most important agent of erosion on our planet, and most of the landscape is sculptured in some way by streams and rivers. As a result, stream valleys are the most widespread landforms on the continents. In the lower reaches of a river, the major landforms are floodplain deposits, deltas, or alluvial fans.

Downcutting of Stream Channels

Downcutting is one of the basic processes of erosion in all stream channels, whether small hillside gullies or the great canyons of major rivers. The process of downcutting is accomplished through the abrasive action of sand, cobbles, and boulders moving along the channel floor. In a very real sense, the sediment load of a river acts like a saw and is capable of cutting down the channel floor at an astonishing rate. Another important factor in the downcutting of a stream channel is the upstream (headward) migration of waterfalls and rapids.

Headward Erosion

In the process of stream erosion and valley evolution, streams have a universal tendency to erode headward, or upslope, and thus to increase the length of the valley until they reach the drainage divide. This process, known as *headward erosion,* is especially significant in a river's upper reaches.

With the universal tendency for headward erosion, the tributaries of one stream can extend upslope and intersect the middle course of another stream, thus diverting the headwater of one stream to the other. This process, known as *stream piracy,* or *stream capture,* is illustrated in Figure 9.1. Stream piracy is most likely to occur if the headward erosion of one stream is favored by a steeper gradient or by a course through more easily eroded rocks.

Slope Retreat

As the drainage network grows larger by headward erosion and deeper by downcutting, the valley walls become subject to a variety of slope processes, such as creep, debris flows, and landslides. These processes, plus the erosive activities of minor tributaries, cause the valley slopes to recede from the stream channel. Slopes can therefore be considered dynamic systems in which the effects of weathering, mass movement, and headward erosion of small gully tributaries combine to transport material down to a main stream. As the system operates, the valley slopes gradually recede from the stream channel, although the valley profile may remain constant.

Careful observation of a landscape will reveal many significant details about the processes of erosion that are presently active, those that were active, and those that may be more important in the future. For example, broad, undissected areas between streams suggest that headward erosion and slope retreat are active processes that will continue until the area is completely dissected. In contrast, sharp, narrow divides indicate that headward erosion is complete and that slope retreat and possible subsequent further downcutting will be more important in the future. Steep stream gradients with numerous rapids and waterfalls indicate that downcutting is a vigorous, ongoing process, whereas gentle gradients with a meandering stream pattern suggest that downcutting is minimal and

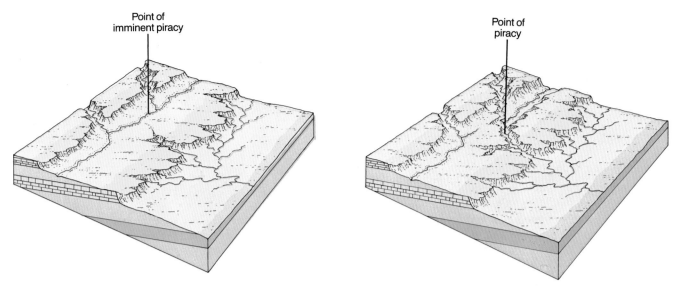

FIGURE 9.1
Stream Piracy occurs where the tributary of one stream erodes headward and intersects the channel of another.

that the stream may be depositing sediment rather than eroding.

PROCESSES OF STREAM DEPOSITION

In the lower part of a river system, the surface of the land slopes gently toward the sea and the stream gradient (the slope of the stream channel measured along the course of the stream) is very low. As a result, the river is unable to transport all of its load, and a significant amount of sediment may be deposited across the floodplain. Sediment deposit across the floodplain occurs through a variety of processes that form point bars, natural levees, and backswamps. A schematic diagram showing features commonly developed on a river floodplain is shown in Figure 9.2.

Much of the sediment load is carried to the sea, however, where it is deposited as a delta. In arid regions, where rivers enter dry basins, the sediment load is commonly deposited as an alluvial fan.

Meanders and Point Bars

All rivers and streams tend to flow in a sinuous path, even if the slope is relatively steep. Once a meander bend is initiated, the flow of water continues to impinge on the outside of the channel, and the bend grows larger and is

accentuated; a small bend thus ultimately grows into a large meander bend (Figure 9.2). On the inside of the meander, velocity is at a minimum, so that some of the sediment load is deposited. These deposits occur on the point of the meander bend and are known as a *point bar*.

As a meander bend becomes accentuated, it develops an almost complete circle. Eventually the river channel cuts across the meander loop and follows a more direct course downslope. The meander cutoff forms a short, but sharp, increase in stream gradient. As a result, the river abandons the old meander loop, which remains as a crescent-shaped lake known as an *oxbow lake*.

Natural Levees

Another key process operating on a floodplain is the development of natural levees. If a river overflows its banks during the flood stage, the water is no longer confined to a channel but flows over the land surface in a broad sheet. This flow pattern significantly reduces the water's velocity, and some of the suspended sediment settles out. The coarsest material is deposited close to the channel, where it builds up an embankment known as a *natural levee*. Natural levees grow with each flood and soon form high embankments. Thus, a river can actually build its channel higher than the surrounding area.

Backswamps

As a result of the growth and development of natural levees, much of the floodplain may be below river level. This area, known as the *backswamp*, is poorly drained and is commonly the site of marshes and swamps. Fine-grained mud, carried by the floodwaters that periodically cover the floodplain, ultimately settles out, adding to the accumulation of floodplain sediment.

Stream Terraces

Many streams may fill part of their valleys with sediment during one period of their history and then erode through the sediment fill during a subsequent period. The factors that may cause a stream to change from deposition to erosion include (1) changes in the volume of discharge as a result of climatic fluctuations, (2) changes in gradient caused by regional uplift, and (3) changes in the amount of sediment load. These fluctuations commonly produce *stream terraces*.

During the last ice age, the hydrology of most rivers changed significantly. Stream runoff was increased greatly by the melting ice, and large quantities of sediment deposited by the glaciers were reworked by the streams, causing many to become overloaded. As a result, many streams that filled part of their valleys with sediment during

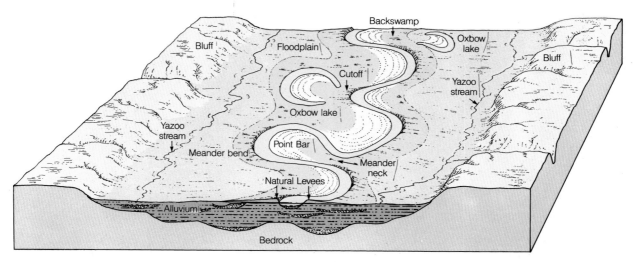

FIGURE 9.2

The Major Features of a Floodplain include meanders, point bars, oxbow lakes, natural levees, backswamps, and small streams. A stream flowing around a meander bend erodes the outside curve and deposits sediment on the inside curve to form a point bar. The meander bend migrates laterally and is ultimately cut off to form an oxbow lake. Natural levees build up the banks of the stream, and backswamps develop on the lower surfaces of the floodplain. Yazoo streams have difficulty entering the main stream because of the high natural levees and thus flow parallel to it for considerable distances before becoming tributaries. Slope retreat continues to widen the low valley, which is partly filled with river sediment.

the ice age are now cutting through that sediment fill to form stream terraces.

Deltas

As a river enters a lake or the ocean, its velocity suddenly diminishes, and most of its sediment load is deposited to form a *delta*. Two major processes are fundamental to the formation and growth of a delta:

1. The splitting of a stream into a distributary channel system, which extends into the open water in a branching pattern.

2. The development of local breaks, or *crevasses,* in natural levees, through which sediment is diverted and deposited as splays in the area between the distributaries.

A major phenomenon in the construction of a delta is the shifting of the entire course of a river. Distributaries cannot extend indefinitely into the ocean because, as the gradient decreases, the river loses its capacity to flow. The river, therefore, is eventually diverted to a new course, which has a

higher gradient. The growth of a delta is also influenced by waves and tides, which are capable of transporting the river sediment farther down the coast or out to sea. The growth of a delta and its general shape are thus largely determined by the balance between the rate of sediment input by the river and the rate of erosion by waves and tides.

Alluvial Fans

An *alluvial fan* is a stream deposit that accumulates in a dry basin at the base of a mountain front. Deposition results from the sudden decrease in velocity as a stream emerges from the steep slopes of the upland and flows across the gentle gradient of the adjacent basin. Alluvial fans form mostly in arid regions where streams flow intermittently.

Although alluvial fans and deltas are somewhat similar, they differ in mode of origin and internal structure. In deltas, stream flow is checked by standing water, and sediment is deposited in a body of water. The level of the ocean or lake effectively forms the upper limit to which the delta can be built. In contrast, an alluvial fan is deposited in a dry basin, and its upper

surface is not limited by water level. The coarse, unweathered, poorly sorted sands and gravels of an alluvial fan also contrast with the fine sand, silt, and mud that predominate in a delta.

PROBLEMS

This exercise involves analyzing a variety of aerial photographs, satellite imagery, and topographic maps of areas where the landscape was formed by stream erosion and deposition. It is an exercise in observation and interpretation. If you have difficulty, review again Exercise 7, which explains how to read maps and photos.

(U.S. Geological Survey) Digital shaded relief map

0   5   10   15 miles

1. Ohio River, Pittsburgh, Pennsylvania

a. What determines the lowest level a tributary stream is able to erode?

b. What percent of the area is not affected by stream erosion?

c. What evidence indicates that the bedrock in this area is horizontal sedimentary rock?

d. Study the patterns of the major tributaries. What evidence suggests that this area has undergone recent uplift?

e. What evidence would support the idea that streams erode (or form) the valleys through which they flow and that the valleys were not formed by some other agent such as faulting, folding, wind, or ice?

f. What geologic process dominates in this area at the present time?

g. How will this area change as the process of stream erosion continues?

   (1) Number of tributaries will increase.

   (2) Length of the tributaries will increase.

   (3) The divides will migrate.

   (4) Downcutting of the tributary stream channel will continue even without crustal uplift.

   (5) Slope retreat will become the dominant process.

h. Study the tributaries and the larger channels into which they flow. In a brief sentence or two state the relationship of the size of the stream and the valley through which it flows.

(U.S. Geological Survey)

CI 20 ft

0   1 mile

2. Kaaterskill, New York

Two drainage systems occur in this area. One flows westward and includes Gooseberry Creek and Schoharie Creek, and the other flows to the southeast and includes Plattekill and Kaaterskill creeks. Striking changes in the drainage have occurred in the past, and more changes will occur in the future. In this exercise, we will examine the reasons why.

a. What is the approximate gradient in feet per mile of Plattekill Creek and Kaaterskill Creek?

b. What is the general gradient in feet per mile of Schoharie Creek and its major tributary, Gooseberry Creek?

c. Will the divide between these two drainage systems migrate to the northwest, southeast, or remain stationary? Explain your answer.

d. What recently happened to the headwater tributaries of Gooseberry and Schoharie creeks?

e. What is the destiny of South Lake and North Lake?

f. Study the drainage pattern of Kaaterskill and Gooseberry creeks and make a series of sketches showing how this drainage has changed.

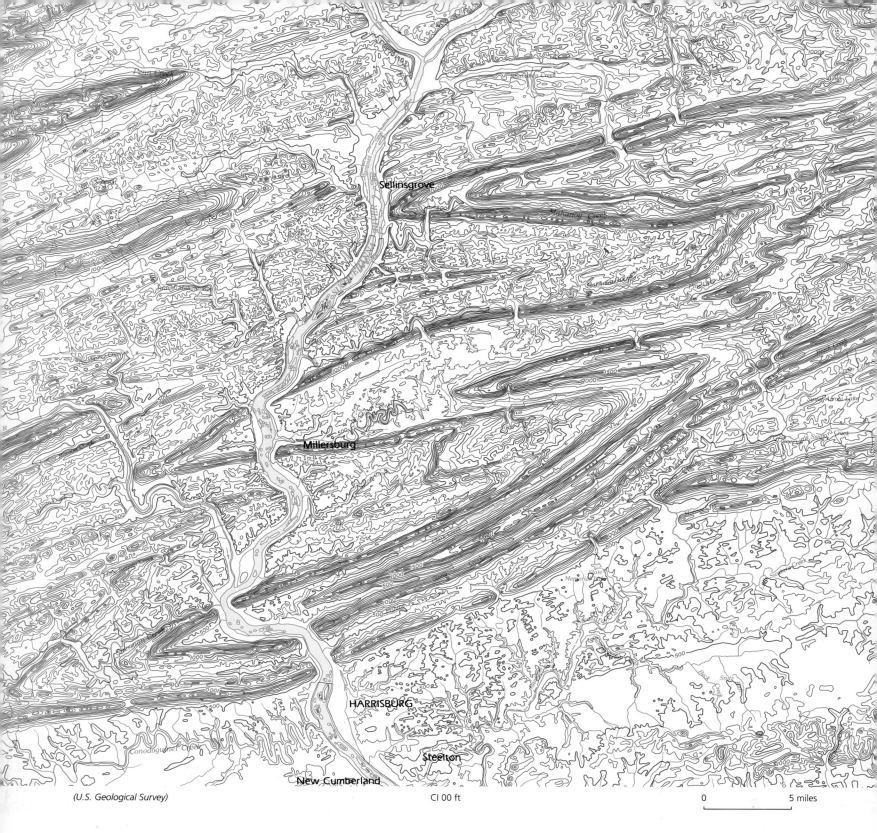

(U.S. Geological Survey)   CI 00 ft   0   5 miles

3. Harrisburg, Pennsylvania

This area is a classic section of the Appalachian Mountains, a belt of strongly folded rocks eroded into long, narrow ridges and intervening valleys. Each ridge is formed on resistant rocks, such as sandstone or conglomerate, whereas valleys are formed on weaker rocks, such as shale. The Susquehanna River has successfully cut across resistant and nonresistant rocks alike and is believed to have been superposed upon the regional structure. Most tributaries to the Susquehanna have developed by headward erosion along nonresistant strata to form strike valleys.

a. Highlight in blue the major tributaries developed by differential erosion along nonresistant strata.

b. With a heavy solid line or in colored pencil, highlight those streams that were superposed. What was the original drainage pattern of the Susquehanna River and its tributaries? What features on the map support your answer?

c. What type of drainage pattern presently exists in this area?

d. Note that there are three fundamentally different streams: (a) the major trunk stream, (b) the major tributaries to the Susquehanna, and (c) tributary streams that flow down the slopes of the mountain ridges. Which stream type is the youngest?

(U.S. Geological Survey) Shaded relief CI 80 ft 0 1 mile

4. Canyonlands, Utah

This area is underlain by horizontal rocks. The meandering pattern of the Green River is typical of that produced by low-gradient streams flowing over a flat floodplain, but here the meandering river flows through narrow canyons 1000 ft deep. The meandering pattern was developed at an earlier stage and was subsequently entrenched as a result of regional uplift.

a. What is the maximum local relief in the area?

b. What is the minimum uplift indicated by the entrenchment of the Green River?

c. Draw a topographic profile along line A–A′.

d. Explain the striking differences between the pattern of the Green River and that of its major tributaries.

e. Study the canyon system cut by the Green River. What major changes have occurred in the course of the Green River since its entrenchment?

f. What changes would you expect to occur in the future?

103

5. Meanders, Alaska

a. Map and label the point bars formed by the major river. Indicate their relative age (1 = oldest, 2 = next oldest, and so forth). How many times has the river changed its course?

b. Express, as a percentage, the amount of shortening of the present course of the major river as compared with the river's course before the cutoff of the large meander shown in the central part of the photograph.

c. What changes would occur in the river gradient as a result of this shortening of the river course when a meander bend is abandoned?

d. What are the most significant geologic processes operating in a meandering river?

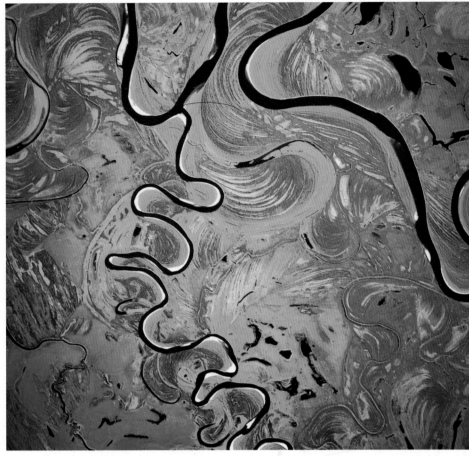

(U.S. Department of Agriculture)

6. Braided Rivers, Nebraska

a. Study the stream pattern and the material between the channels. Is the sediment deposited by present stream processes?

b. How do the processes that form a braided river differ from those that form a meandering system?

c. What factors influence the development of a braided river pattern?

(U.S. Department of Agriculture)

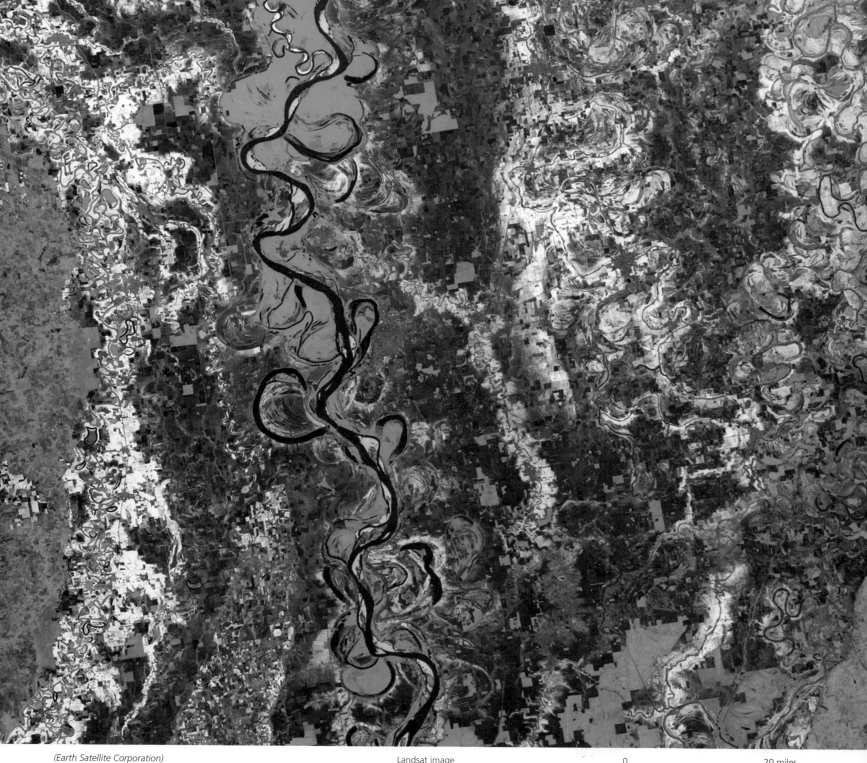

(Earth Satellite Corporation)　　　　　　　　Landsat image　　　　　　　　0 ⊢——————————⊣ 20 miles

7. Mississippi River

This is a Landsat image of the Mississippi River in Arkansas and Mississippi. Most of the area is covered with floodplain sediment, but bedrock underlying the upland adjacent to the floodplain is expressed by distinctive vegetation (shown as dark red).

a. Identify this area on the physiographic map on pages 94–95. Estimate the width of the floodplain.

b. Locate the areas where the Mississippi River has changed its course significantly in recent years. Describe the process by which these changes occur.

c. What landforms dominate this area?

d. Study the oxbow lakes along the Mississippi River and make a series of sketches to show how an oxbow lake changes with time.

e. Where are the youngest oxbow lakes?

f. Where will the next oxbow cut-off likely occur?

105

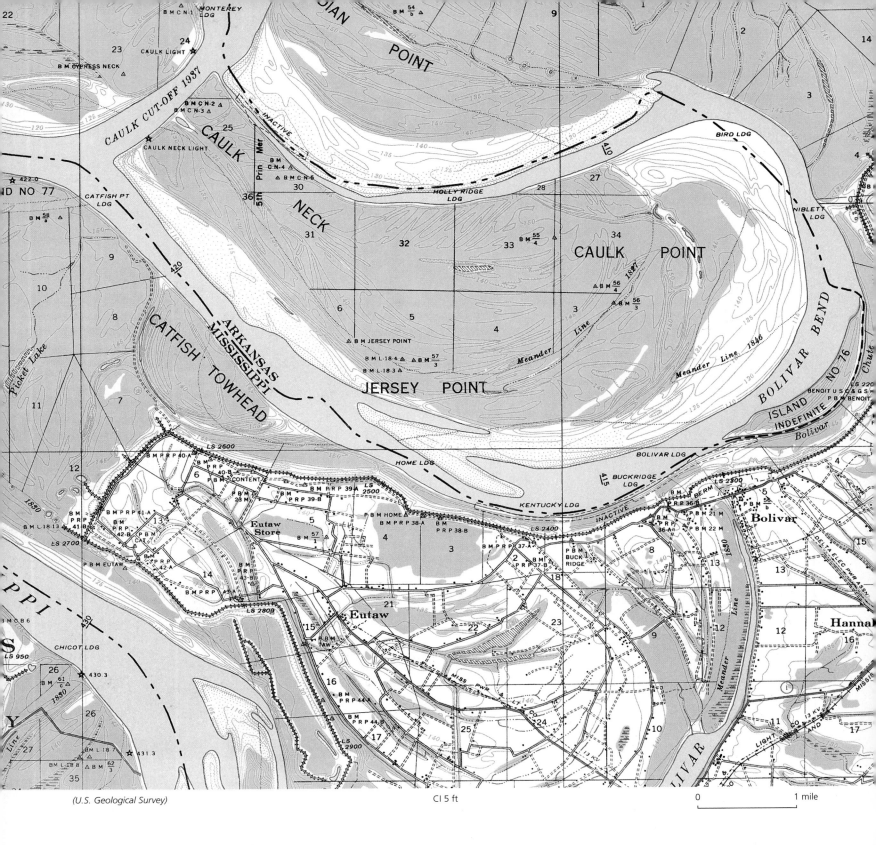

(U.S. Geological Survey)   CI 5 ft

0   1 mile

8. Lamont, Arkansas

This is a map of topographic detail of part of the Mississippi River and the adjacent floodplain shown on the preceding page. Note that the contour interval is only 5 ft. Study the present course of the river and previous courses established by surveys during historic times (shown on the map by a dotted line labeled "meander line" and the date).

a. What major change occurred in this area in 1937?

b. What major changes occurred between 1827 and 1937?

c. What is the maximum local relief of the natural levees in this area?

d. Color the oldest meander scars green and the youngest yellow.

e. If the Mississippi River is 40 ft deep in this area, estimate the volume of sediment deposited on a point bar each year. Note the position of the meander line in 1827, 1846, and the position at the time of cutoff in 1937.

Landsat image

0   1 mile

9. New Orleans, Louisiana

Note that the land subdivision is *not* the typical north–south grid. Instead, land boundaries are perpendicular to the river because the high, dry ground on the natural levee is much more valuable as farmland than is the wet backswamp. The patterns of farmland therefore outline the natural levees (light blue).

a. Study the pattern of land adjacent to the river and outline the boundaries of the natural levees.

Use a soft lead pencil at first; when you are satisfied with your results, trace the boundary with a colored pencil or felt-tipped pen. How wide are the natural levees?

b. What geologic processes operate in the floodplain? What indications of these processes are apparent on this image?

c. How wide are the natural levees near the left margin of the image?

d. What evidence suggests that the Mississippi River has significantly changed its course?

e. What is the origin of Lake Pon-chartrain?

f. How are natural levees expressed at the lower end of the Mississippi River, below New Orleans?

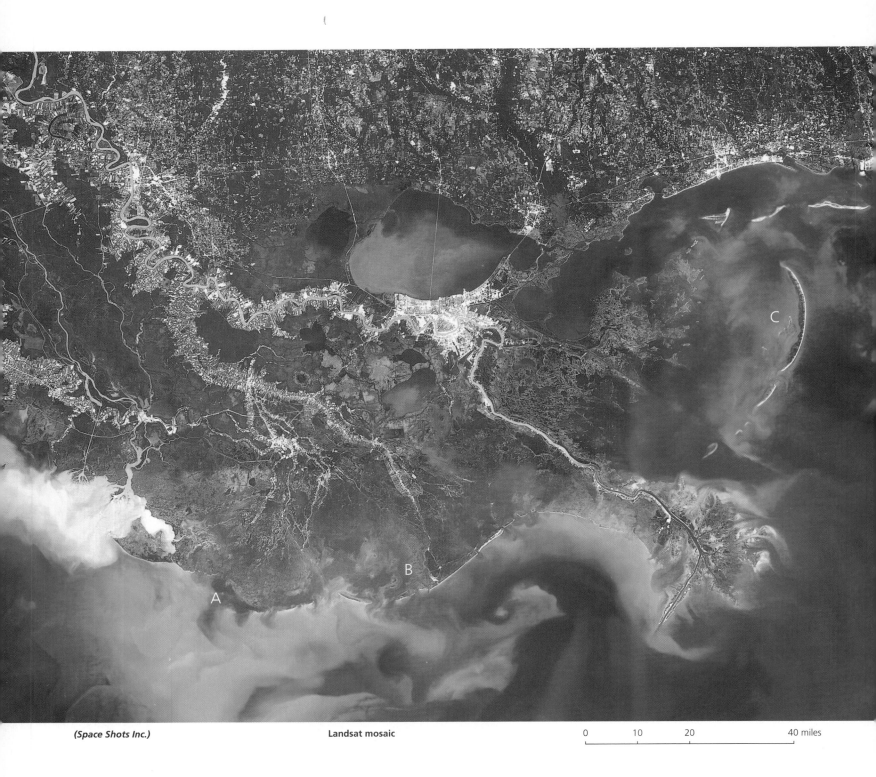

(Space Shots Inc.)   Landsat mosaic

0   10   20   40 miles

10. Mississippi Delta

a. What evidence indicates that a major distributary once existed in the areas of points A, B, and C?

b. Where was the point of avulsion for the development of each of these distributaries?

c. How did the barrier islands east of point C form?

d. What geologic processes operate on a subdelta after the main river changes its channel to a new course?

e. Why are there no beaches or barrier bars near the present active

subdelta, whereas they are common on older abandoned distributaries?

f. How are natural levees expressed on this satellite image?

g. In which direction is most of the suspended sediment moving in the waters offshore?

11. Alluvial Fans—California

a. What evidence do you find in this area that substantiates the assumption that streams cut the valleys through which they flow?

b. Sediment of different ages can be recognized on the fans by differences in tone, texture, and relative elevation. Map these sediments according to their relative age (1 = oldest sediments, 2 = next oldest, and so forth).

c. Why do the drainage channels on the fan split into numerous distributaries?

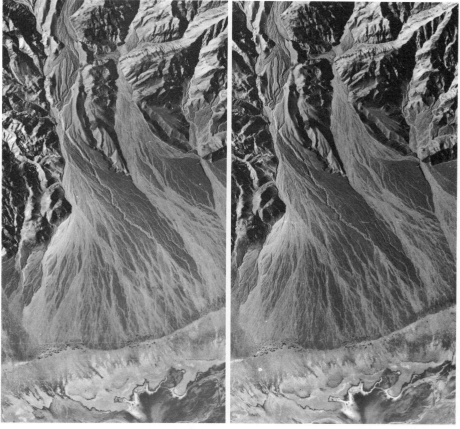

(U.S. Department of Agriculture)   Stereogram   0 _____ 1 mile

12. Stream Terraces—Montana

a. What processes were involved in the deposition of the sediment on which the terraces are cut?

b. Color the stream terraces according to their relative age (yellow = oldest, green = next oldest, and blue = youngest). What geologic events do the terraces record?

c. Draw an idealized cross section across the stream valley and show the morphology of the terraces and their relationship to the bedrock below.

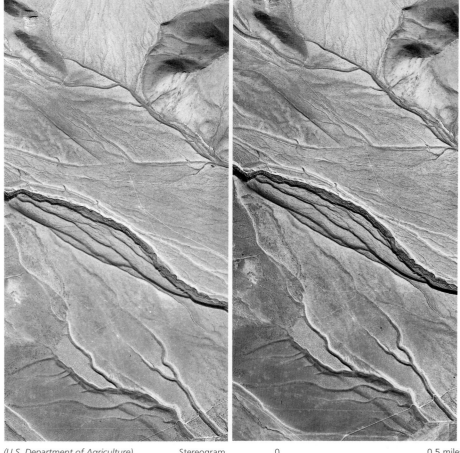

(U.S. Department of Agriculture)   Stereogram   0 _____ 0.5 miles

(Earth Satellite Corporation)

Landsat image

0   30 miles

13. Death Valley, California

Death Valley is part of the Basin and Range province, which is characterized by fault-block mountains and down-dropped basins partly filled with sediment derived from erosion of the adjacent ranges. (See the physiographic map on pages 94–95.) It is an arid region (summer temperatures commonly rise above 120 °F) of internal drainage.

a. Carefully study the image. How many fans (large or small) can you see?

b. What factors govern the size of an alluvial fan (mountain height, climate, size of the drainage system)?

c. Draw a line showing the location of the fault along which the mountain was uplifted. Is there any evidence that recent movement has displaced some of the fans?

d. How would erosion and deposition in this area change if the climate became more humid?

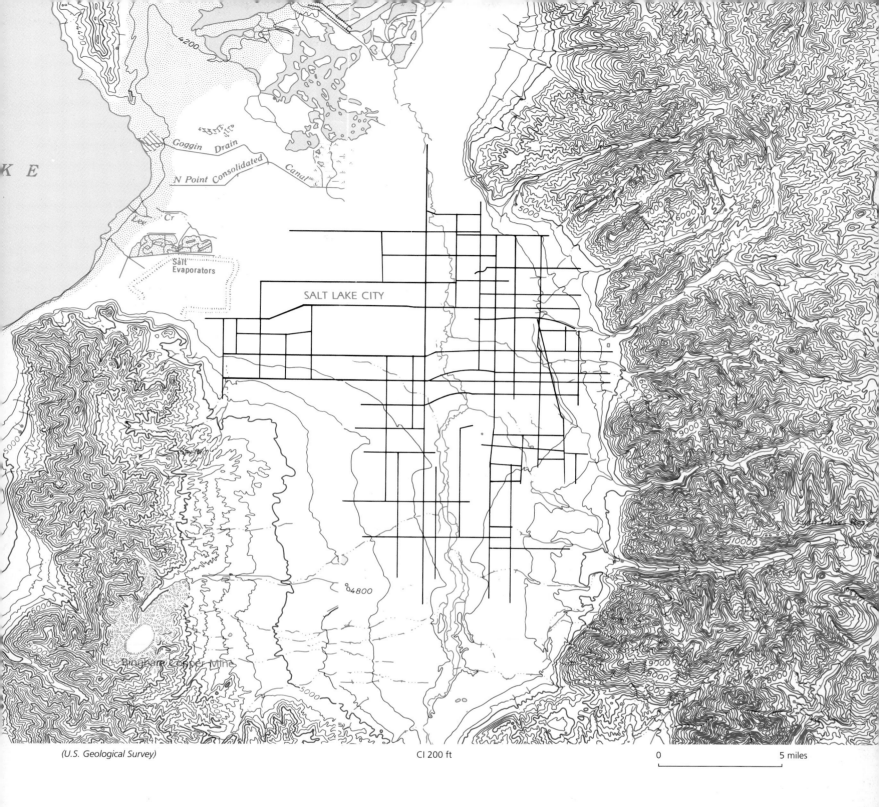

(U.S. Geological Survey)   CI 200 ft   0   5 miles

14. Salt Lake City, Utah

Salt Lake City is also in the Basin and Range province. The Jordan River is the only permanent stream in the area. It flows northward through the central part of the valley and collects drainage from adjacent mountains. Most of the streets and cultural features of Salt Lake City have been eliminated from this map to enable you to study the landforms and drainage system better.

This area has experienced serious flooding from the rapid melting of heavy snow coupled with heavy spring rain.

a. Study the contour lines in the Salt Lake Valley and, where necessary, emphasize them with a brown pencil. On the basis of the contour lines and the drainage system coming from the mountains, trace the path of the drainageways as they flowed across the valley prior to urbanization.

b. On the map, show those areas of the city that would be most susceptible to flooding.

c. One of the flood-control measures used was to sandbag the streets to create channels in which runoff flowed. With a blue line, show on the map those streets that would serve best as sandbagged channels for flood runoff.

d. Outline in red those areas that could be considered at risk from landslides and mud flows.

MASS MOVEMENT

OBJECTIVE

To recognize the distinctive surface features produced by mass movement and to understand the role mass movement plays in the evolution of slope systems.

MAIN CONCEPT

The movement of surface material downslope is a universal erosional process. It occurs on the steep slopes of high mountains, on gentle rolling plains, on sea cliffs, and on the slopes of the ocean floor. Mass movement is thus a significant factor in slope erosion and slope retreat.

SUPPORTING IDEAS

1. The most important factors influencing mass movement are (a) the saturation of slope material with water, (b) freezing and thawing, (c) the oversteepening of slopes, and (d) earthquakes.

2. The most significant types of mass movement that produce landforms large enough to be seen on aerial photographs and topographic maps, are (a) rock falls, (b) debris flows, (c) landslides, and (d) rock glaciers.

Mass movement includes all types of slope failure. Because of its potential for destruction, mass movement has been studied by engineers as well as geologists. As a result, it has been classified in various ways. In this exercise, we will consider four main types that produce features large enough to be seen on aerial photographs and topographic maps.

ROCK FALLS

Rock falls include the free fall of fragments, ranging from a single grain to huge blocks weighing millions of tons. Small-to-moderately sized fragments are commonly broken loose by ice wedging and accumulate at the base of a cliff as a *talus cone*.

DEBRIS FLOWS

A *debris flow* is a mixture of rock fragments, mud, and water that flows downslope as a viscous fluid. Movement can range from a flow that is similar to that of freshly mixed concrete to that of a fluid stream of mud with a velocity nearly equal to that of running water. Debris flows include *mud flows* composed mostly of fine-grained material,

coarse debris involving huge block of rock, and *lahars* (volcanic mud flows).

LANDSLIDES

Although the vague term *landslide* has been applied to almost any kind of slope failure, a true landslide involves movement along a well-defined slippage plane. A landslide block moves as a unit along a definite fracture, with much of the material moving as a large slump block. The detached block leaves behind a distinct curved incision, or scar. The slippage plane is typically spoon shaped. As the block moves downward and outward, it commonly rotates in such a way that bedding or other identifiable surfaces are tilted backward toward the slippage plain. In the lower part of the slump block, part of the displaced material may move as a debris flow. The characteristic scar, tilting of bedding or other surfaces, and jumbled, poorly drained hillocks formed by previous slides serve to identify terrains that have been modified by landslides.

ROCK GLACIERS

Rock glaciers are long, tongue-like masses of angular rock debris with pore

spaces filled with ice. They resemble a glacier in general outline and form. The surface of a rock glacier is typically furrowed by a series of parallel flow ridges, similar to those in an advancing lava flow. Evidence of movement includes concentric wrinkle ridges, a lobate form, and a steep front. Measurements indicate that rock glaciers move as a body downslope at rates ranging from 2 in./day to 3 ft/yr.

Rock glaciers commonly occur at the heads of glaciated valleys and are fed by a continuous supply of rock fragments produced by ice wedging on the cirque wall. Ice in the pore spaces between the rock fragments presumably is responsible for much of the flow movement. With a continuous supply of rock fragments from above, the weight of the mass increases and causes the mass of ice and rock to flow. Favorable conditions for the development of rock glaciers thus include steep cliffs and a cold climate.

PROBLEMS

Answer the questions associated with the photographs on pages 113–115.

1. San Juan Mountains, Colorado

a. List the types of mass movement that you can identify on this photograph.

b. One type of mass movement in this area forms large, tongue-like landforms with a wrinkled upper surface. Map these features and explain how they form.

c. What types of mass movement probably occur in this area but do not produce a major landform?

d. List the geologic conditions in this area that are conducive to mass movement.

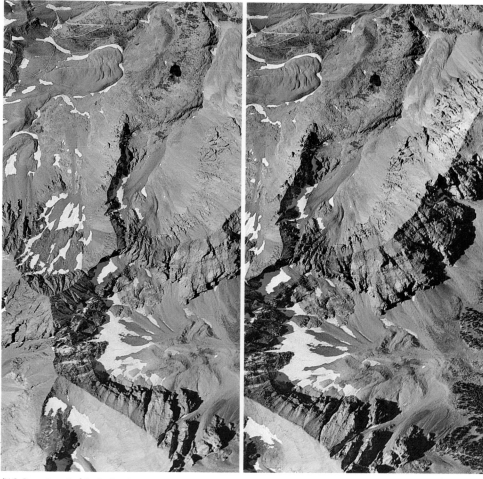

(U.S. Department of Agriculture)

0   1 mile

2. Mass Movement—British Columbia, Canada

a. What types of mass movement have occurred in this area?

b. What natural geologic conditions exist in this area that would be conducive to mass movement?

c. What effect did this pulse of mass movement have on the geologic processes operating in the valley and on human activity?

d. What evidence suggests that there have been several pulses of movement in the "landslide" areas?

e. Compare and contrast the mass movement in this area with that shown in the photograph above. How are they similar?

(Department of Energy, Mines, and Resources, Ottawa, Canada)

0   0.5 miles

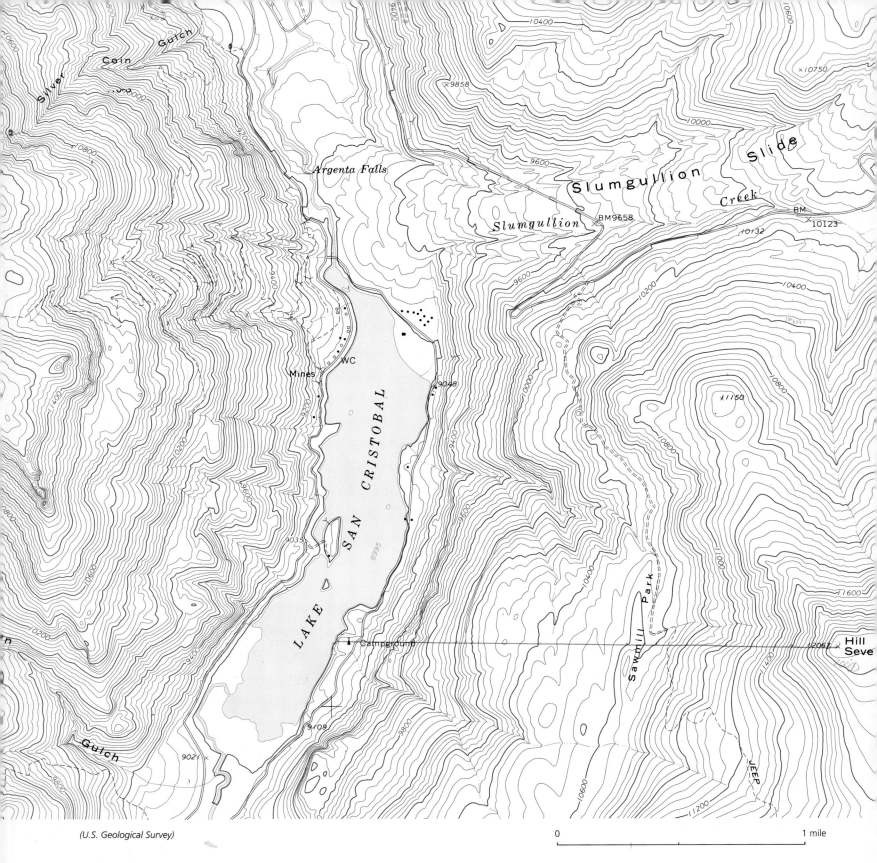

(U.S. Geological Survey)

0   1 mile

3. Lake San Cristobal, Colorado

This area is in the San Juan Mountains of Colorado. The rocks are mostly Tertiary volcanics, with andesites and pyroclastic rocks dominating.

a. What is the origin of Lake San Cristobal?

b. What type of mass movement is the Slumgullion slide?

c. What material most likely is involved in the Slumgullion slide?

d. How thick is the slide debris near the shores of Lake San Cristobal?

e. What was the approximate depth of the lake when it was first formed?

f. Trace the boundaries of the Slumgullion slide on the map, and draw an arrow pointing toward the source, or head, of the slide.

g. What areas might be considered a high risk for future slides? Show these areas on the map with a red felt-tipped pen.

(U.S. Department of Agriculture)

0   1 mile

4. Mass Movement—New Mexico

The small mesa in this area is capped by a layer of basalt. The rock beneath is shale.

a. What type of mass movement occurs in this area? List evidence to support your answer.

b. What geologic factors are necessary for this type of movement to develop?

c. What will eventually happen to the mesa?

d. How is the mass movement debris removed from the base of the mesa?

OBJECTIVE

To recognize the distinctive landforms produced by groundwater and to understand how landscapes are formed by solution activity.

MAIN CONCEPT

The solution activity of groundwater is a significant agent of erosion in regions underlain by soluble rocks such as limestone, gypsum, and rock salt. Where solution activity dominates, it commonly develops a distinctive landscape, *karst topography,* which is characterized by sinkholes, solution valleys, disappearing streams and, in tropical regions, by residual conical hills.

SUPPORTING IDEAS

1. Karst topography in a given region commonly evolves through a series of stages until the soluble rock is completely removed.
2. The style of karst topography depends on climate and the characteristics of the bedrock.
3. Tower karst is a distinctive type of karst topography common in humid climates.

KARST TOPOGRAPHY

Karst topography develops in humid climates where there is abundant water free to circulate through a soluble limestone terrain. Arid regions are not susceptible to extensive karst development because of low rainfall, and in the cold regions of the arctic and subarctic, groundwater is permanently frozen and not free to circulate. Therefore, tropical, humid regions provide the ideal combination of abundant water and high temperature to enhance the chemical reactions that drive the solution process conducive to the development of karst topography.

Landforms of Karst Topography

The most characteristic landforms of a karst terrain are closed depressions, or *sinks,* or *sinkholes,* that range from small pits a few feet in diameter to large basins more than a mile wide. Two main mechanisms are involved in the development of sinks: (1) the roof collapse of near-surface caves and (2) accelerated solution and subsidence in areas of closely spaced fractures.

A second major landform in karst regions is a *solution valley,* which typically forms where sinks enlarge and merge into a single elongate depression. As a result, a karst region as a whole lacks a well-integrated surface drainage system. Tributaries are few and are generally very short. Many minor streams appear suddenly as springs in blind valleys, flow for a short distance, and then disappear into sinkholes, so that much of the drainage is underground. Only major streams flow in defined, open valleys.

Karst in Humid Temperate Regions

An idealized sequence of stages in the evolution of a karst topography in temperate regions is shown in Figure 11.1. In the early stage (Figure 11.1A), solution activity forms a system of underground caverns that enlarge until, eventually, the cavern roof collapses, thus producing a sinkhole. As the sinkholes increase in number and size, some merge to form solution valleys. When solution valleys become numerous and interconnect, the area is considered to have reached the intermediate stage of development (Figure 11.1B). Continued solution activity removes most of the limestone, so that only scattered, rounded hills and knolls remain into the late stage (Figure 11.1C).

Karst in Semiarid Regions

In arid to semiarid regions, there is a limited amount of water to circulate in the subsurface, so that karst topography is rare or poorly developed. Sinks are typically small and widely separated, and are commonly associated with short disappearing streams.

Tower Karst

The most spectacular type of karst topography is developed in humid tropical regions where abundant rainfall and high temperatures promote rapid solution activity of the surface and subsurface water. The general landscape of tower karst is dominated by residual hills, some more than 1000 ft high.

A. Early Stage: (1) Surface is nearly flat with a few small, scattered sinkhole depressions. Subterranean caverns are numerous. (2) Throughout the early stage, sinkholes become more abundant and increase in size.

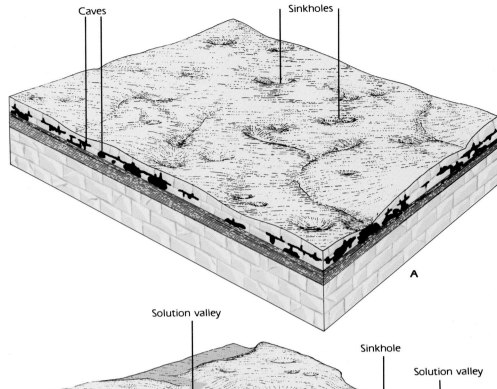

B. Intermediate Stage: (1) Individual sinkholes enlarge and merge to form solution valleys with irregular branching outlines. (2) Much of the original surface is destroyed. There are many springs and disappearing streams. (3) Maximum relief, although not great, is achieved. Differences in elevation between the rim and floor of a sinkhole rarely exceed 200 to 300 ft.

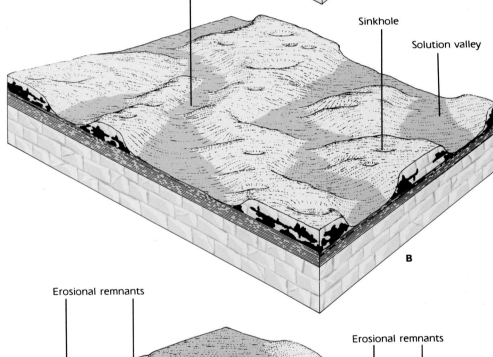

C. Late Stage: (1) Solution activity has reduced the area to the base of the limestone unit. (2) Hills formed as erosional remnants are few, widely scattered, and generally reduced to low, conical knolls.

FIGURE 11.1
Idealized Evolution of a
Karst Topography

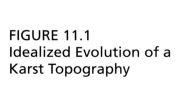

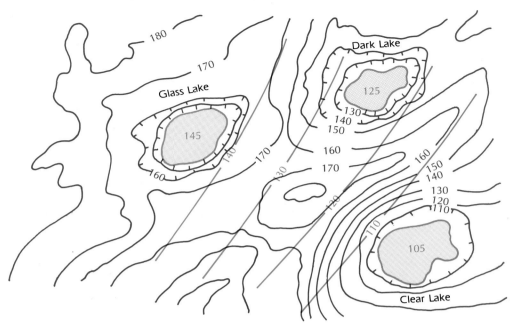

FIGURE 11.2
Contouring a Water Table
In the figure, brown lines represent surface contours and blue lines are contours on the surface of the water table.

Contouring a Water Table

The elevation of the lakes in a region provides important information about the groundwater conditions. The surface of each lake is essentially the surface of the water table. Because the lakes are control points for the elevation of the water table, we can construct a generalized contour map showing the configuration of the water table from those data.

For example, on the map in Figure 11.2, the elevation of Dark Lake is between 120 and 130 ft, the elevation of Glass Lake is between 140 and 150 ft, and the elevation of Clear Lake is between 100 and 110 ft. For convenience, assume that the lake levels are 125, 145, and 105 ft, respectively.

The water table can be contoured by applying the same principles used in contouring the land surface. The 140- and 130-ft contour lines would be located between Dark Lake and Glass Lake, and the 120- and 110-ft contour lines would be located between Clear Lake and Dark Lake. We thus observe that the water table slopes to the east.

PROBLEMS

There are three main parts to this exercise: (1) analyzing various types of karst topography, (2) studying groundwater and its effect on the environment, and (3) determining the configuration of the water table. If you do not live in an area of karst topography, you will be introduced to a new and fascinating landscape and the processes responsible for its formation.

1. West Texas

a. Compare the sinkholes in this area with those in Puerto Rico (the photographs below) and with those in the Mammoth Cave area (photographs on the next page). Consider such features as number, size, shape, and depth.

b. If the areas in all three figures are underlain by limestone, why is the topography of each so different?

c. Trace the drainage system. How does it differ from the drainage in the area shown in the photographs below?

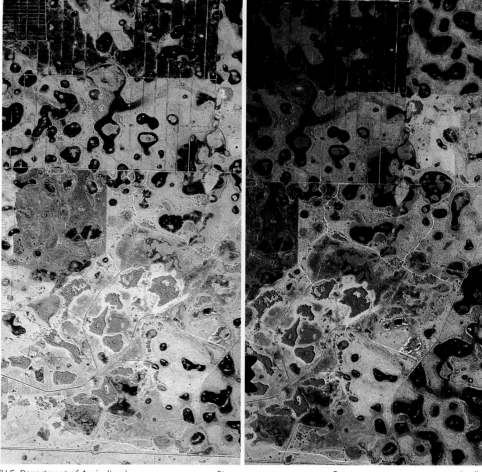

(U.S. Department of Agriculture)   Stereogram   0   1 mile

2. Manati, Puerto Rico

The sinkholes in this region are as much as 150 ft deep, and the hills are up to 300 ft high. This is an area of classic tower karst, in which the residual hills have steep sides or are nearly vertical. Many hills are conical, whereas others form elongate ridges. The sinks are star-shaped, elongate, or hemispheric depressions. One is so nearly a perfect hemisphere in cross section that it has been used as the foundation for a giant radio telescope.

a. List the evidence indicating that this topography was *not* produced by stream erosion.

b. Explain the origin of the elongated hills in this region.

c. Study the area shown in stereo and map the major drainage patterns.

d. How does this type of karst topography differ from that in the Mammoth Cave area (photographs on the next page)?

e. What major factors control the development of tower karst?

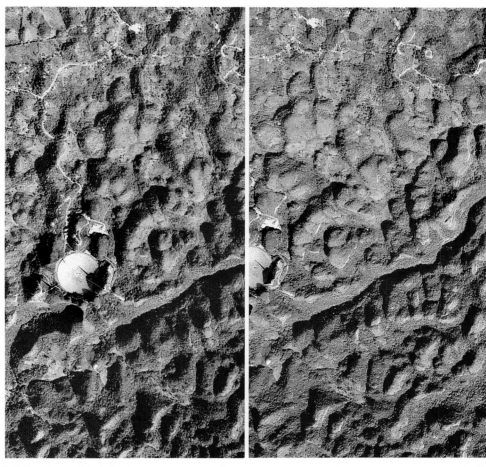

(U.S. Department of Agriculture)   Stereogram   0   1 mile

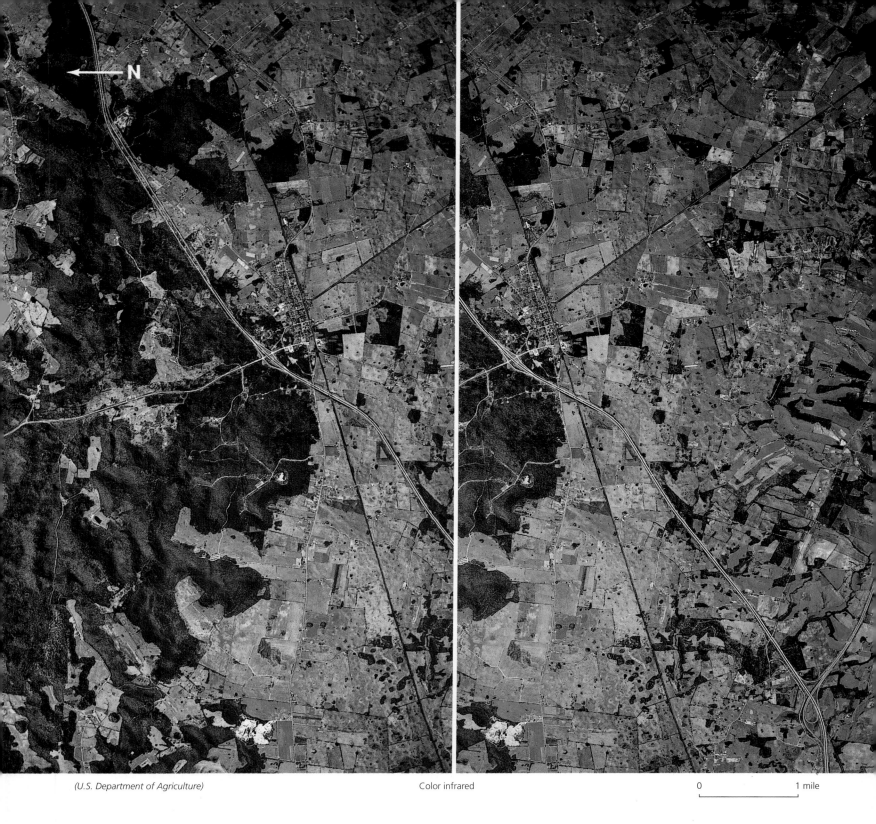

(U.S. Department of Agriculture)　　　　Color infrared　　　　0　　　1 mile

3.　Mammoth Cave, Kentucky

The pockmarked surface of the Mammoth Cave area of Kentucky is a classic example of karst topography. Many sinks are aligned along joint systems and thus form linear depressions. Some sinks south of Park City are more than 60 ft deep. North of Park City, the number of sinks per unit area diminishes, but a vast underground network of caverns is here, including Mammoth Cave.

a. Study the area carefully and outline the large solution valleys.

b. Locate a stream and trace its course on the photo. Does this stream show characteristics of a typical river system (a network of branching tributaries, a trunk stream, or a valley system, for example)? What features of karst drainage patterns are unique?

c. Is there evidence to indicate that the development of sinkholes is an ongoing process, or did they form during one particular period of time?

d. How does a sinkhole change with time?

e. What geologic hazards are most likely to be encountered in this area?

f. What particular problems of waste disposal and pollution does a karst area present to a city or to industrial development?

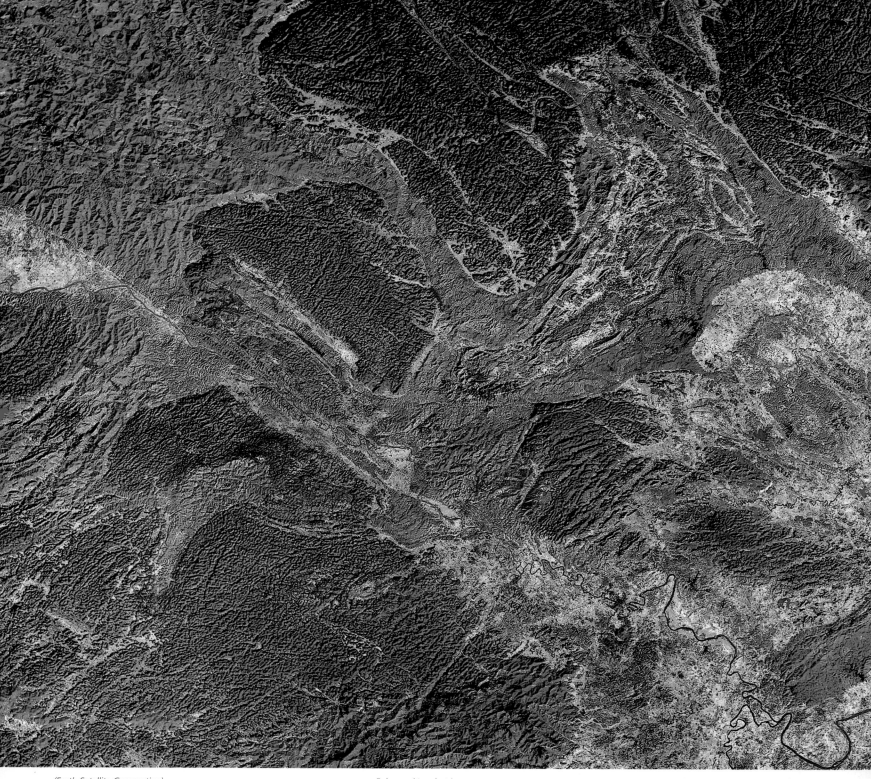

Enhanced Landsat image

0   5 miles

4. Nanming, China

This image shows China's Nanming karst area. Although local details are not apparent on satellite imagery, the regional texture of the surface is exceptionally well expressed.

a. What karst landforms (sinkholes, solution valleys, or erosional remnants, for example) dominate the area?

b. What evidence indicates that a fracture system in the bedrock has influenced this area's karst topography?

c. Would the karst landforms in this area be more similar to those of Kentucky, Puerto Rico, Florida, or Texas? Explain the basis for your answer.

5. Southern Florida

In recent years, the rapid urban development of Florida's Dade County has posed a serious threat to the environment, despite the efforts of some residents to maintain environmental quality. Most of the land surface in south Dade County is fewer than 10 ft above high-tide level, and the highest areas are only 25 ft above sea level. Water is the primary environmental concern here. For this reason, Dade County and its adjacent counties are part of a regional water control system developed to conserve and protect both the surface water and groundwater and to control freshwater and saltwater flooding. The water-control measures include management of the drainage canals as well as water conservation and management of storage areas.

The major drainageways shown on this Landsat image include patterns of water movement in natural drainageways, the Everglades Swamp, and canals. The canals form a system designed to reduce flood damage from storms and tides and to conserve fresh water.

A limestone sequence underlying the area contains two aquifers. The upper one, the Biscayne Aquifer, extends from the surface to a depth of 120 ft and is the source of the area's fresh water. The other aquifer, the Floridian Aquifer, is more than 800 ft below the surface, and contains only brackish and salt water.

Water problems now being faced by south Dade County are saltwater intrusion into the Biscayne Aquifer, contamination of the aquifer by recharge from canal and surface waters, and depletion of the aquifer. The demand for water to serve the residents of the community has lowered the water table in the Everglades and threatens the natural flora and fauna there.

In an area of such low-lying topography, drainage problems are serious as well. Flooding is always a potential hazard. Drainage canals have helped to control the flooding, but the canals adversely affect the ecosystem of the Everglades by diverting the swamp's freshwater source. To prevent drainage problems in urban areas, building codes that require a minimum elevation for lots and streets have been established. The consequent storm runoff, which results from urbanization, increases the flooding from the high tides caused by hurricanes.

South Dade County faces the environmental problems that plague all urban areas, plus the problems created by the county's particular physical characteristics. Changes are occurring rapidly, as the natural forests and the Everglades are being converted to agricultural use and eventually to commercial and residential development.

a. Use a series of arrows to show the original surface drainage patterns in southern Florida on the photograph.

b. What effect has the construction of canals had on this drainage?

c. What areas would be most susceptible to damage from hurricanes and heavy rainfall?

d. What effect would the Florida Keys have on hurricanes approaching from offshore?

e. What particular problems need to be considered before building on a limestone terrain?

f. Why has most of the urbanization of this area been concentrated in its present location?

g. How would you expect the pattern of urbanization to appear 20 years hence?

h. What would be the best sites for the disposal of solid waste? Why?

i. What problems are likely to result from excessive pumping of groundwater?

Enhanced Landsat image

0   10 miles

(Earth Satellite Corporation)

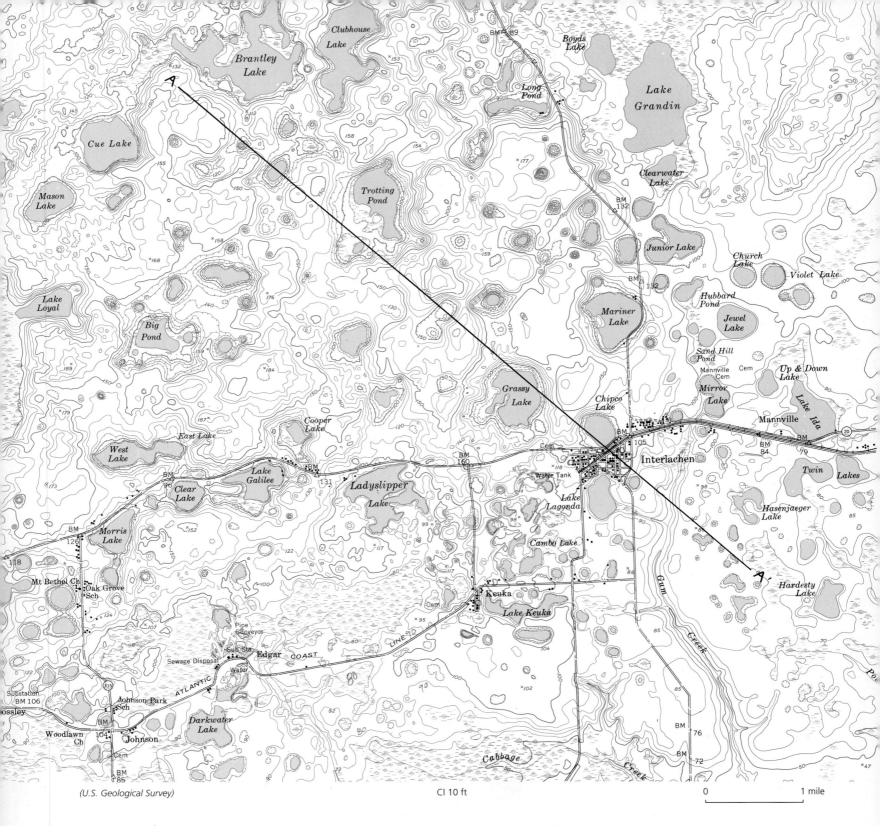

(U.S. Geological Survey) CI 10 ft 0   1 mile

6. Southern Florida

a. Determine the elevation of most of the large lakes in this area, and construct a contour map of the surface of the water table. (Refer again to Figure 11.2 and the discussion of how to contour the surface of a water table.) Use CI = 10 ft.

b. Construct a topographic profile along line A–A'. Draw the water table with a blue line. Be sure to show how Gum Creek influences the water table.

c. Use a series of arrows to indicate the general direction of groundwater movement on the map.

d. How deep would you have to drill to obtain water if the well site were located at the road intersection 1 mi west of Interlachen?

e. Many farms, small industries, and urban centers dump all of their untreated liquid waste into the subsurface. What happens to this waste? In what direction would the waste contaminants move in this area?

f. What natural hazards (floods, landslides, subsidence, earthquakes, or erosion) have the most geologic significance in terms of influencing construction work in this area?

OBJECTIVE

To recognize the types of landforms developed by valley glaciers and to understand the processes responsible for their development.

MAIN CONCEPT

Valley or alpine glaciers are systems of moving ice that flow down preexisting stream valleys. As the ice moves, it erodes the landscape, creating deep U-shaped valleys, sharp interstream divides, cirques, horns, and hanging valleys. Sediment deposited near the end of the glacier typically forms terminal moraines, part of which is reworked by meltwater and deposited as outwash.

SUPPORTING IDEAS

1. During a glacial epoch, many thousands of miles of ice flow through systems of valley glaciers and erode the former stream valleys by abrasion and glacial plucking.
2. Ice wedging and mass movement along the valley walls deliver large amounts of rock debris to the glacier. The debris is then transported as lateral and medial moraines.
3. Deposition occurs at the end of the glacier, where the rate of melting exceeds the rate of ice flow. Terminal moraines and outwash plains are major depositional features.

GLACIAL SYSTEMS

A glacier is an open system of flowing ice. Water enters the system as snow, which is transformed into ice by compaction and recrystallization. The ice then flows through the system under the pressure of its own weight and leaves the system by evaporation and melting. The balance between the rate of accumulation and the rate of melting determines the size of the glacial system (the body of ice).

The essential parts of a glacial system are (1) the *zone of accumulation,* where there is a net gain of ice, and (2) the *zone of ablation,* where ice leaves the system by melting, calving, and evaporating. In the zone of accumulation, the direction of movement is inclined downward with respect to the surface of the glacier. The degree of downward inclination decreases from the head of the glacier to the snow line. At the snow line, the direction of movement is parallel to the surface of the glacier. In the zone of ablation, the movement is upward toward the surface, with upward inclination increasing from the snow line to the snout.

Extending and Compressing Flow

The movement of glacial ice is not uniform. The velocities of ice flow in the zone of accumulation increase progressively from head to the snow line. Here, the ice is under tension and is constantly pulling away from up-valley ice. This is the condition of *extending flow.* Below the snow line, velocities progressively decrease, so up-valley ice is continually pushing against down-valley ice. This is a condition of *compressing flow.*

Where bedrock slopes steepen, glacier velocities increase and extend-ing flow prevails; where the bedrock slopes are gentle, velocities decrease and compressing flow occurs (Figure 12.1). Where glaciers descend over extremely steep slopes, the ice descends with high velocities, creating a veritable icefall, in slow motion. These are zones of extreme extending flow, and the ice is greatly thinned and completely crevassed. The flow velocity in an icefall can exceed ten times that of the glacier elsewhere along its course. At the base of an icefall, conditions are reversed; flow decreases rapidly, the compressing flow dominates, and the glacier thickens.

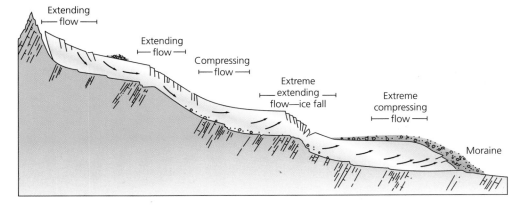

FIGURE 12.1
Extending and Compressing Flow

Crevasses

The most obvious and abundant structures in a glacier are crevasses—large open cracks formed by fracturing a brittle upper layer of ice as the underlying ice continues to flow. Crevasses are nearly vertical and may be more than 30 meters deep and thousands of meters long. Crevasses are tensional fractures produced by differential motion in the ice. Almost any part of a glacier involved in differential flow velocities is likely to develop crevasses transverse to the direction of ice flow. Chevron crevasses are present in almost all valley glaciers, along their lateral margins, as a result of stretching produced by differences in flow velocity as the ice drags along the valley walls (Figure 12.2). These crevasses are usually short and point upstream. Transverse crevasses form at right angles to the direction of flow where flexing of the ice occurs as the glacier moves over bumps or ridges on the bedrock floor. Similarly, icefalls are intensely crevassed by the greatly accelerated rate of flow as the ice moves down a steep slope. Longitudinal crevasses develop at the terminus of a glacier, where the ice stream spreads out, setting up tensional stresses at right angles to the flow direction.

VALLEY GLACIERS

The diagrams in Figure 12.3 illustrate how valley glaciers modify a landscape previously sculptured by running water. Before glaciation (Figure 12.3A), the topography is characterized by V-shaped stream valleys and rounded hills. In looking up the major valleys, we see that the spurs (or ridges) appear to overlap because the stream naturally curves and bends. Valley size is proportional to the size of the stream that flows through it. At stream junctions, tributaries join the major streams without waterfalls or rapids.

During glaciation (Figure 12.3B), snowfields form in the high mountain ranges, and glacier systems expand down the major stream valleys. Glacial erosion, together with ice wedging, tends to produce sharp, angular landforms. Deep erosional depressions,

called *cirques,* develop at the glacier head and erode headward. Where several cirques merge, a sharp, angular peak, called a *horn,* is produced. Rock fragments derived from erosion of the valley walls and mass movement accumulate on the margins of the glacier and are dragged along by the moving ice, forming ridges of debris, called *lateral moraines*. Where two ice streams join, the lateral moraines merge in the middle of the combined ice streams to form a long, narrow ridge called a *medial moraine.* Lateral and medial moraines are an expression of the flow lines of the glacier and do not cross each other but may merge near the glacier terminus.

When the glacial period ends and the ice melts (Figure 12.3C), the topography is strikingly different from the preglacial landscape. Horns and cirques dominate the landscape in the high mountain regions and form some of the most scenic landscapes in the world. Cirques are semicircular in plan view and are often described as resembling a bowl or an amphitheater. They commonly contain small lakes and even small vestiges of glacial ice. Divides between glacial valleys develop sharp ridges known as *arêtes.* Valleys that have been glaciated are typically straight and U-shaped in cross section. Major glacial valleys are deep, and spurs are truncated. Hanging valleys, which occur where glacial tributaries enter the larger glacial valleys, often have spectacular waterfalls. Glacial val-

leys cut below sea level are often invaded by the sea to become fiords.

Near the end of the glacier (not shown in the diagrams in Figure 12.3), the amount of erosional debris transported by the ice increases greatly, until at the very end of the glacier there may be more sediment than ice, and the top of the glacier may be completely covered with rock debris. This material constitutes the glacier's *terminal moraine.* When the ice melts, the terminal moraine characteristically extends as a broad arc, conforming to the shape of the terminus of the ice. A terminal moraine commonly will trap meltwater from the receding glacier to form a temporary lake. If periods of stability occur during the recession of the ice, a *recessional moraine* can form behind the terminus. The large volume of meltwater released at the terminus of a glacier reworks much of the previously deposited moraine and redeposits the material in an *outwash plain* beyond the glacier.

PROBLEMS

This exercise is designed to show the dynamics of glaciation and the spectacular landscape formed during the last ice age. We will study maps and photographs where glaciers are still active and areas where glaciers have recently disappeared. Your best visual reference is Figure 12.3. If you understand these diagrams, you will enjoy working on this exercise.

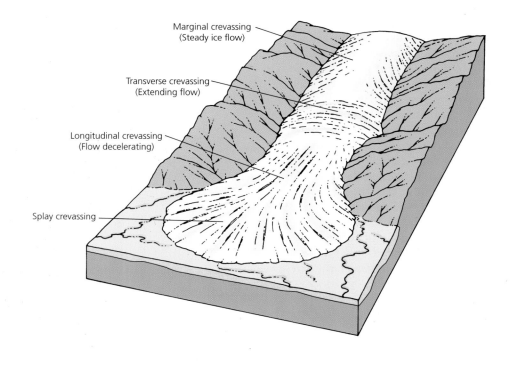

FIGURE 12.2
Types of Crevasses

A. Preglaciation: Topography before glaciation is characterized by V-shaped stream valleys, overlapping ridges, and rounded hills.

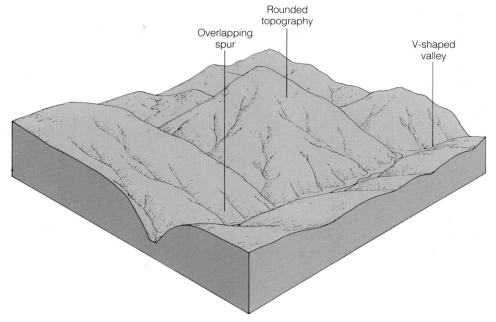

B. Glaciation: Valley glaciers develop from snowfields in the high peaks and expand down stream valleys. Major glaciers thus have a network of tributaries that follows the preexisting drainage system.

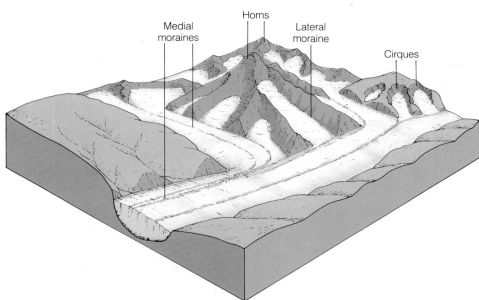

C. Postglaciation: Broad, deep, U-shaped valleys are the most characteristic landform developed by valley glaciation. Cirques, horns, and arêtes are glacial features that create spectacular scenery in the highlands. Hanging valleys, often with high waterfalls, occur where tributaries enter the main valley.

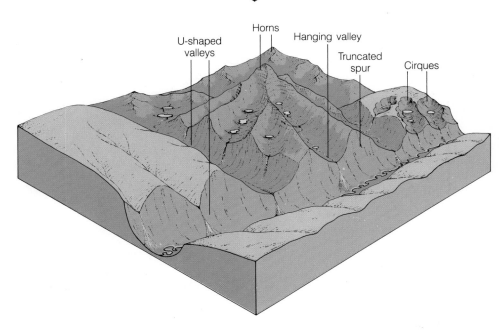

FIGURE 12.3
Landforms Developed
by Valley Glaciation

1. Valley Glaciers—Alaska

a. What is the source of the material that makes up a moraine?

b. Are the moraines moving or are they stationary?

c. What is the origin of a medial moraine?

d. At least fourteen medial moraines occur in the main glacier. How many tributary glaciers occur in this system?

e. What evidence in this area indicates that glacial ice flows?

f. Study the ice in the small tributaries. Does it flow faster, slower, or at about the same velocity as the ice in the main stream?

g. Study the patterns of the crevasses and indicate, on the photograph, areas of major extending flow.

h. Label the major horns, arêtes, and cirques on the photograph.

(U.S. Department of Agriculture) Color infrared 0 1 mile

2. Valley Glaciers—Alaska

a. What type of drainage pattern occurs in the outwash plain?

b. Why does this pattern develop?

c. What is the origin of the numerous small lakes in the terminal and recessional moraines?

d. What evidence indicates that at one time the glacier extended beyond the "terminal" moraine?

e. How many recessional moraines can you identify? How do they form?

f. Compare and contrast the nature of sediment in the moraines with that in the outwash plain.

g. What evidence indicates that the terminus of a glacier is a zone of compressing flow?

(U.S. Department of Agriculture) Color infrared 0 1 mile

3. Valley Glaciers—Alaska

a. What evidence indicates that lateral and medial moraines are derived from mass movement on the valley walls as well as from glacial plucking?

b. What evidence indicates that the various tributary glaciers flow at different rates?

c. Explain the difference in the color and texture of the medial moraines.

d. What evidence shown in these photographs indicates that glacial ice flows?

e. What causes the crevasses in the tributary glaciers where they enter the main stream?

(U.S. Department of Agriculture) Color infrared stereogram 0 1 mile

4. Valley Glaciers—Alaska

a. Several different terminal moraines are distinguishable by differences in form, texture, and growth of vegetation (red color). Outline the major terminal moraines in the photographs and explain briefly the glacial history they record.

b. Explain why the end of this glacier is completely covered with moraine sediments.

c. How can you distinguish between older and younger moraines in this area?

d. What is the origin of the small lakes near the bottom of the photographs?

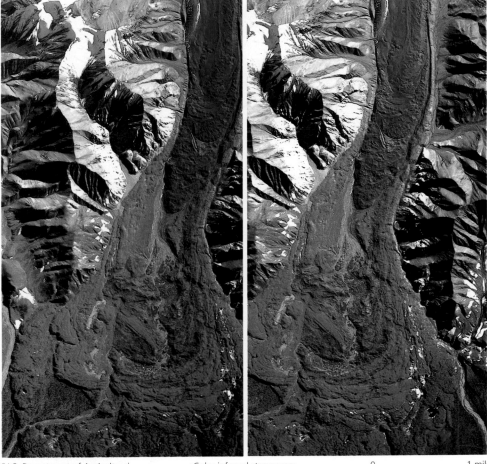

(U.S. Department of Agriculture) Color infrared stereogram 0 1 mile

129

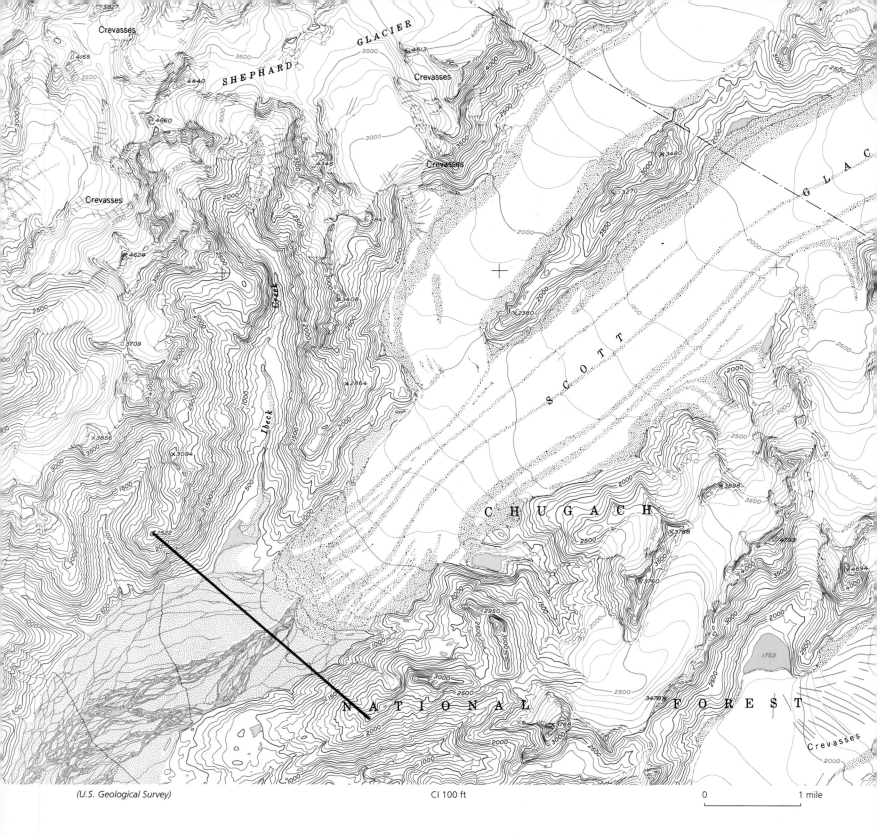

(U.S. Geological Survey)   CI 100 ft   0   1 mile

5. Cordova, Alaska

a. Study the landforms and label the following features: (1) outwash plain, (2) arêtes, (3) recessional moraines, (4) medial moraines, (5) hanging valleys, (6) horns, (7) braided streams, (8) cirques, and (9) ice falls.

b. Draw arrows on the map to show the flow direction of the tributaries to the Scott Glacier.

c. Draw a topographic profile across the valley approximately 1 mi southwest of the end of Scott Glacier. How does a glacial valley differ from a stream valley?

d. Can we reasonably assume that a glacier modifies the valleys through which it flows? What field evidence can you cite to support your answer?

e. Draw a longitudinal profile up Scott Valley to the front of the glac-

ier. If the gradient were projected up the valley under the glacier, what would be the approximate thickness of the ice at the 1800-ft contour line near the middle of Scott Glacier?

f. Study the blue contour lines showing the surface of the glaciers. Indicate, on the map, areas of extending flow (E) and compressing flow (C).

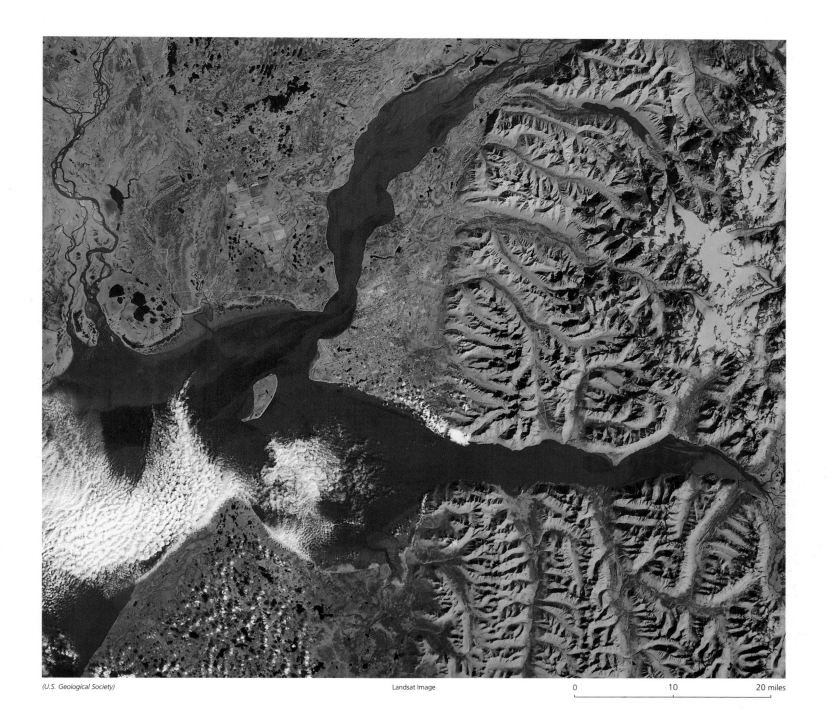

(U.S. Geological Society) Landsat Image

0   10   20 miles

6. Glacial Topography— Anchorage, Alaska

During the ice age, extensive snowfields and alpine glaciers were prevalent in the high mountain ranges of the world. Many of these glaciers have now disappeared, but they have left a clear depositional and erosional record of their former activity. Careful study of glaciated landscapes permits us to outline the areas of former glaciers and to interpret the Pleistocene history of the regions.

a. Identify and label several examples of the following: (1) horns, (2) glacial valleys, (3) arêtes, (4) outwash plains, and (5) fiords.

b. Explain the origin of the lake in the upper right part of the image.

c. Explain why the main stream is braided.

d. What is the origin of the numerous small lakes in the lower center of the image?

e. Explain why there are no moraines in this area.

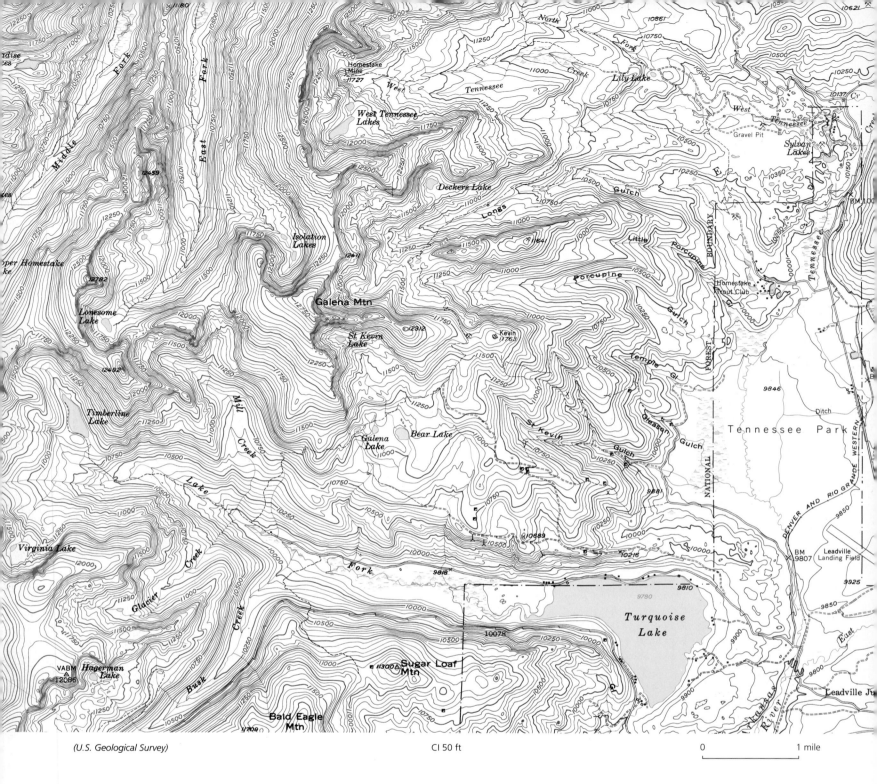

(U.S. Geological Survey) CI 50 ft 0   1 mile

7. Holy Cross, Colorado

a. Study carefully the landforms on the map and color those areas in blue that you believe were formerly covered by glaciers. (A good approach is to first outline the major cirques and then trace the margins of the U-shaped valleys downstream.)

b. Study the landforms near the lower end of the major valleys. Locate the moraines and color them green.

c. Compare the glacial features in this area with those in the photograph on page 131. List the glacial features common to both areas and then list those that are unique to the Holy Cross area.

d. What is the origin of the small lakes located at the head of many of the major valleys?

e. Why have swamps and lakes developed in the central part of many of the major valleys?

(U.S. Geological Survey)

Landsat

0   1 mile

8. Denali National Park, Alaska

a. A valley glacier system is similar to a river system in that it consists of a major trunk stream and a series of tributaries. How many glacier systems are still active in this area?

b. Draw a line between the zone of accumulation and the zone of ablation. Does this line approximate a contour line? Why?

c. What features indicate that glaciers were once more extensive than at present?

d. Show, with a series of arrows, the direction of ice movement in areas that are not covered with ice at the present time.

e. List the features formed in the outwash plains that you recognize in this image.

EXERCISE 13 CONTINENTAL GLACIATION

◼ OBJECTIVE

To recognize the major types of landforms created by continental glaciers and to understand the processes responsible for their development.

◼ MAIN CONCEPT

A continental glacier is a system of flowing ice, up to 2 mi thick, that covers large parts of a continent. As the ice within the glacier moves, it erodes the soil and bedrock and transports the erosional debris towards the margins of the ice, where the debris is deposited in a variety of landforms.

◼ SUPPORTING IDEAS

1. Much of the sediment load of continental glaciers is deposited at the ice margins to form end moraines.
2. Drumlins, eskers, and ground moraines form beneath the ice.
3. Meltwater from the glacier may rework large amounts of glacial debris and deposit it in an outwash plain in front of the ice sheet.
4. The last ice age created most of the surface features in the Great Lakes region and adjacent areas.

Continental glaciers originate in polar regions where precipitation is sufficient to build up and maintain a thick body of ice. Their margins are commonly irregular or lobate because the advancing ice front moves more rapidly into lowlands than over higher terrain. The thickness of a continental glacier commonly exceeds 10,000 ft but is rarely more than 15,000 ft because the strength of ice cannot support any greater thickness. If more ice is added by greater amounts of precipitation, the ice will not become thicker; it will simply flow faster.

An important effect of continental glaciation is that the weight of the ice depresses Earth's crust, so that the land surrounding the ice sheet slopes toward the glacier. This sloping of the terrain toward the ice sheet may produce lakes along the ice margins or it may temporarily permit an arm of the sea to invade and fill the depression.

The idealized diagrams in Figure 13.1 illustrate the major landforms developed by continental glaciation and the processes by which they are produced. A number of significant landforms result from the geologic processes operating at the margins of continental glaciers. It is there that the flowing ice leaves the glacial system by melting and evaporation (Figure 13.1A).

Much of the sediment carried by the glacier is deposited along the ice margins as moraine or is reworked by meltwater and deposited as outwash. The sediment deposited at the ice margins follows the lobate outline of the glacier to form terminal moraines, recessional moraines, and interlobate moraines. Lakes that form in the depressions next to the ice sheet margins receive sediment from the meltwater streams. This sediment accumulates as deltas or may be dispersed over the lake bottom.

Elsewhere, meltwater issues from tunnels beneath the ice and forms braided streams that deposit stratified fluvial sand and gravel on the outwash plain. Blocks of ice broken from the glacier may be partly buried by sediment as the glacier begins to recede. Moraine and outwash sediment may be reshaped by subsequent advances of the glacier to produce streamlined hills, known as *drumlins*.

When the glacier retreats, ridges of moraines mark the former positions of the ice margins, and well-sorted, stratified clay and silt mark the sites of former lakes (Figure 13.1B). The outwash plain is characterized by deposits of fluvial sand and gravel. Depressions, known as *kettles,* form where partly buried ice blocks have melted. Sinuous ridges, or *eskers,* result from sediment deposited on the floor of former ice tunnels, and numerous lakes fill depressions in the ground moraine. Preglacial drainage has been greatly modified or obliterated, so that the postglacial drainage is completely deranged.

PROBLEMS

Much of the landscape of the northern United States and essentially all of Canada was sculptured by continental glaciers. In this exercise, you will study local areas on topographic maps and aerial photographs, and regional features on satellite imagery. Try to relate each map and photograph to the glacial map shown on pages 140–141. This map was compiled from hundreds of man-years of field study and summarizes our current understanding of glacial features in the Great Lakes region. If you understand this map, you can begin to appreciate the tremendous impact continental glaciation had on Earth's surface.

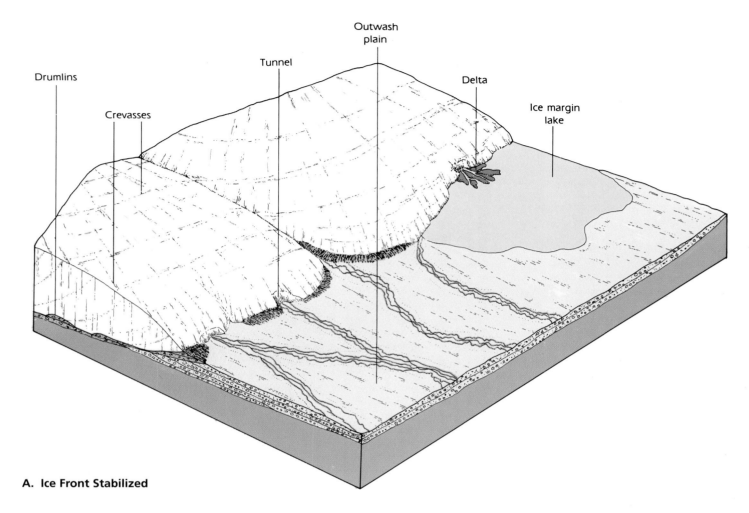

A. Ice Front Stabilized

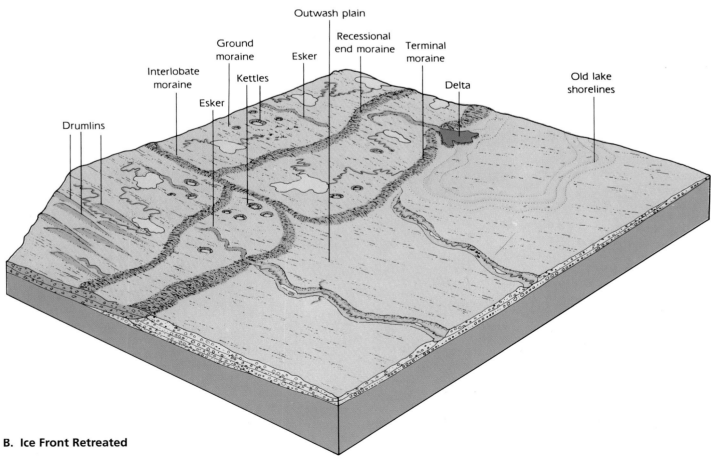

B. Ice Front Retreated

FIGURE 13.1
Landforms Developed by Continental Glaciation

135

1. Drumlins—Canada

a. Consider the shape, orientation, and grouping of the drumlins shown here. Show, with arrows, the direction of glacier movement.

b. Cite evidence to support the conclusion that drumlins result from glaciation and not from some other geologic process, such as wind action or catastrophic flooding.

c. To what extent has stream erosion modified the drumlins?

d. Locate the low, sinuous ridges trending roughly parallel to the drumlins and color them yellow. Explain how these features originated.

e. Many cities use sanitary landfills to dispose of solid waste. What problems might arise if this were the method of waste disposal in a drumlin area?

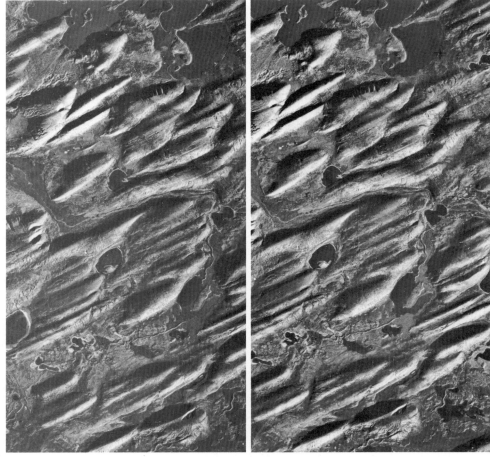

(Department of Energy, Mines, and Resources, Ottawa, Canada)

0   1 mile

2. Glacial Landforms—North Dakota

a. Draw a line along the boundary between the end moraine and ground moraine. What topographic features characterize each area?

b. What is the origin of the lakes in the northeastern part of the area?

c. Draw several arrows on the photograph to show the direction of ice movement. Explain how you reached this conclusion.

d. Why is the drainage so poorly developed in this area?

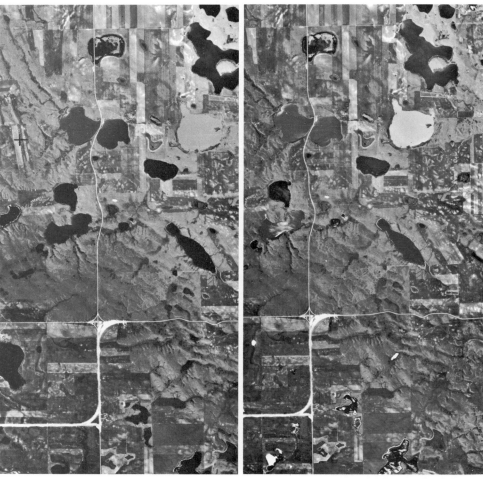

(U.S. Department of Agriculture)

0   1 mile

(U.S. Geological Survey)　　　　　　　　　　　　　CI 10 ft　　　　　　　　　　0　　　　　　1 mile

3. Jackson, Michigan

a. Locate this area on the map on page 141 (Jackson, Michigan, is located in the south-central part of the state.) Study the landforms in this area and identify (a) eskers, (b) ground moraine, and (c) terminal moraine. Draw the approximate boundaries between these features. Color the ground moraine light green, the terminal moraine dark green, and the eskers orange.

b. The poorly developed drainage and the extensive swamp and marsh are typical of areas where continental glaciation has occurred. To emphasize the lack of integration in this drainage system, use a dark-blue pencil to trace the drainage of the Grand River and its tributaries. How does this drainage pattern compare with the drainage patterns of the river systems studied in Exercise 9?

c. Is there any evidence from the Jackson drainage pattern that the landforms are primarily depositional and that they have been only slightly modified by erosion?

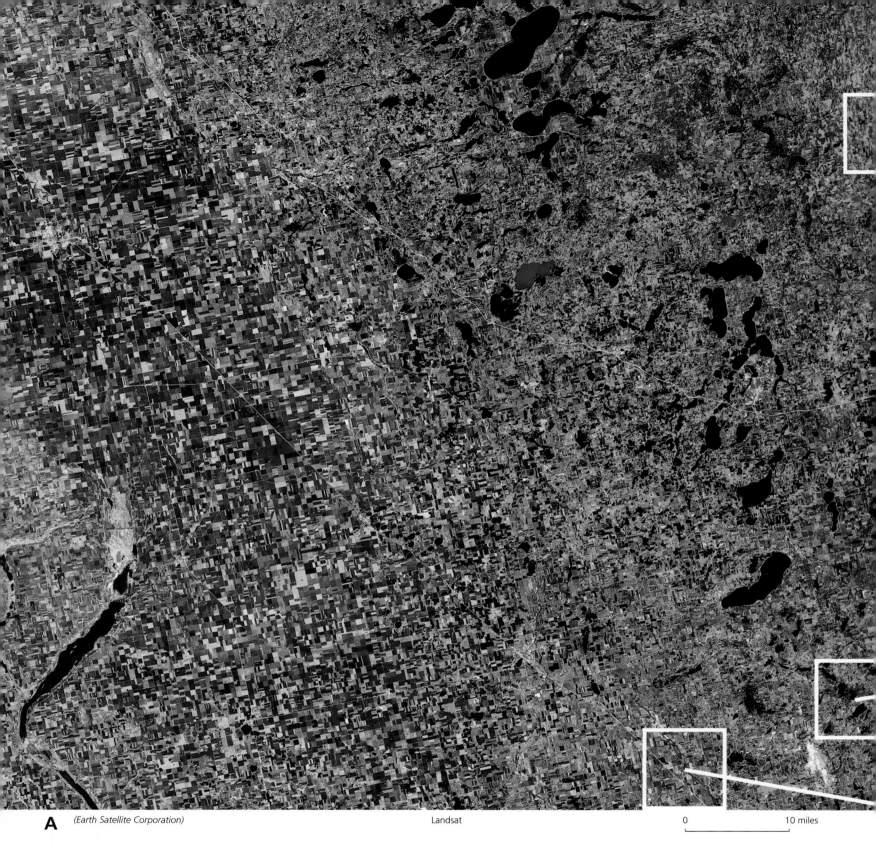

A *(Earth Satellite Corporation)*   Landsat   0   10 miles

4. Glacial Topography— North Dakota

a. On the Landsat image, draw a line along the contacts between lake sediment and end moraine, and between end moraine and ground moraine. What topographic features characterize each area?

b. What is the origin of the numerous small lakes shown in the upper central part of the image?

c. What is the nature and origin of the linear fabric shown in the extreme northeast corner of the Landsat image? Study also the high-altitude, color infrared photograph in (B).

d. Describe the topography shown in the high-altitude photograph in (C). How is this general topography apparent on the Landsat image (A)?

e. Describe the topography shown in (D). How does it differ from that shown in (C)?

f. On the basis of the glacial land-forms shown in (A), construct a generalized paleogeographic map outlining the former position of the ice front. Use arrows to indicate the direction of ice movement.

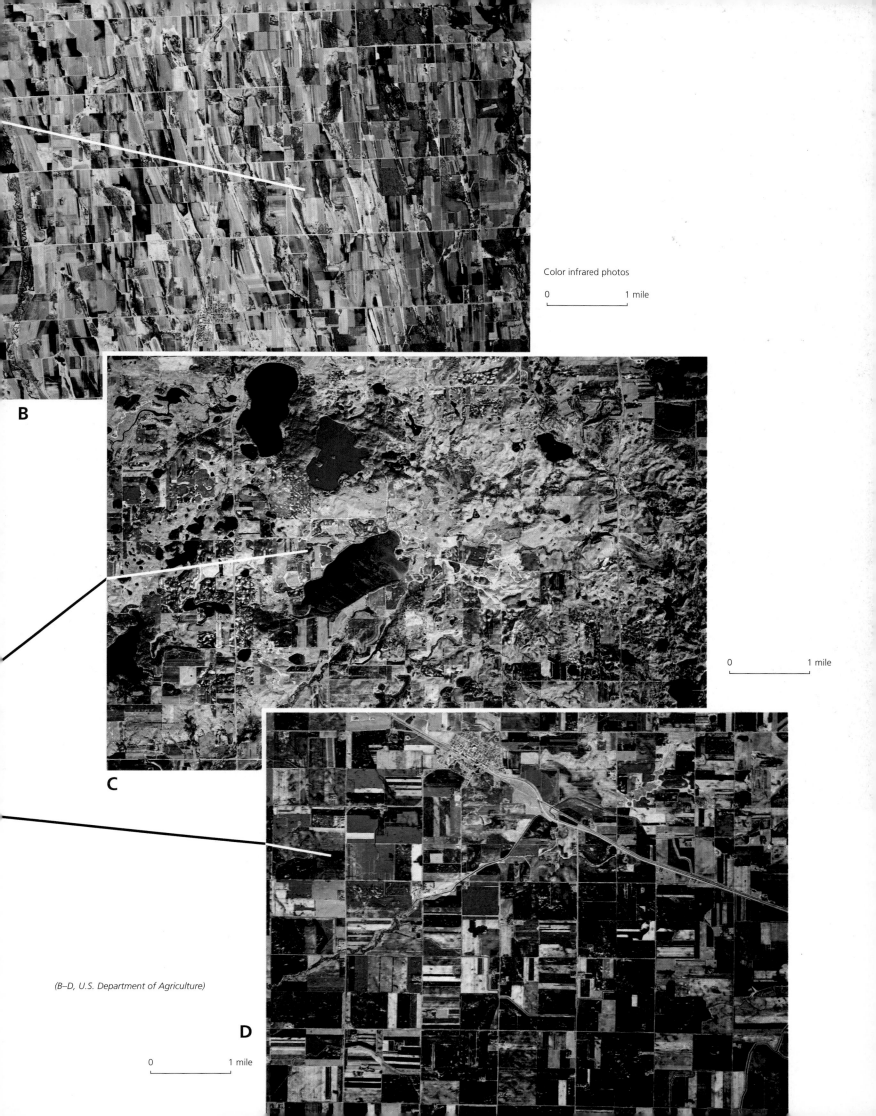

Color infrared photos

0   1 mile

B

0   1 mile

C

(B–D, U.S. Department of Agriculture)

D

0   1 mile

5. Glacial Features—Great Lakes Area

a. Study the patterns of the major glacial features (outwash, lakes, and moraines), and note the contact of the older and younger deposits associated with them. On the basis of these relationships, determine the relative ages of the following: ground moraine or drift, eskers, outwash sediments, and lake sediments.

b. How many glacial stillstands are represented by the end moraines in Ohio and Indiana, southwest of Lake Erie? How many stillstands are represented by the moraines southwest of Lake Michigan?

c. Was the glacial front advancing or retreating when these moraines formed?

d. Compare and contrast the glacial features formed by the lobes of ice that advanced down the areas of Lake Erie, Lake Michigan, and Green Bay into eastern Wisconsin.

e. Note the breaks in many of the end moraines. What is the probable cause of these breaks?

f. In the northeastern Indiana area, study the pattern of outwash sediments (yellow) in the vicinity of the moraines near Union City, Mississinewa, Salamonie, Wabash, and Fort Wayne. What does this pattern indicate about the direction of the drainage during glacial times?

g. What effect did the end moraines have on the location of ancient glacial lakes such as lakes Maumee, Whittlesey, Glenwood, and Calumet (indicated by red lines on the map as lacustrine sediments and structures)? These ancient lakes extended beyond the present-day shores of Lake Michigan and Lake Erie.

SYMBOLS

Strandlines

Striation direction

Streamline features

(Geological Society of America)   0   25 miles

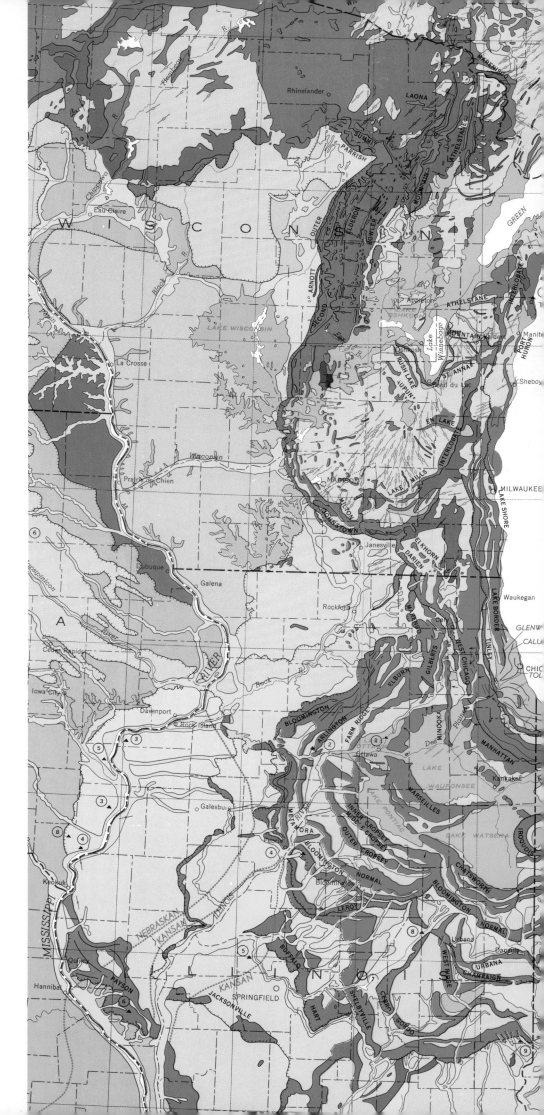

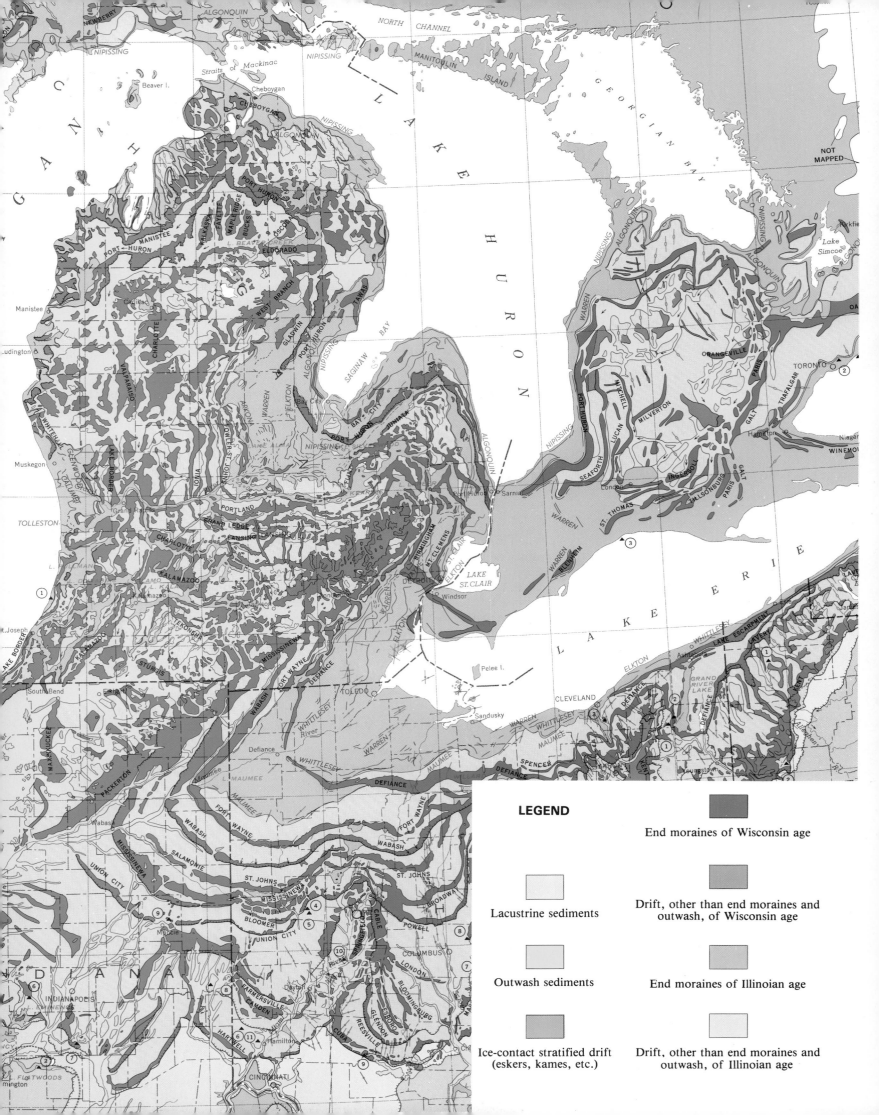

LEGEND

End moraines of Wisconsin age

Lacustrine sediments

Drift, other than end moraines and outwash, of Wisconsin age

Outwash sediments

End moraines of Illinoian age

Ice-contact stratified drift (eskers, kames, etc.)

Drift, other than end moraines and outwash, of Illinoian age

OBJECTIVE

To recognize the major types of coasts and to understand the processes involved in their development.

MAIN CONCEPT

The configuration of a shoreline changes as a result of erosion and deposition until it reaches a state of equilibrium.

SUPPORTING IDEAS

1. Wave refraction is a fundamental process by which energy is concentrated on headlands and dispersed across bays.

2. Longshore drift is one of the most important processes of sediment transport along a coast.

3. Erosion along a coast tends to develop sea cliffs by the undercutting action of waves and longshore currents. As the cliff recedes, a wave-cut platform develops.

4. Most coasts can be classified on the basis of the geologic processes that were most significant in determining their configuration.

5. The worldwide rise in sea level, associated with the melting of the continental glaciers, drowned many coasts. The configuration of our present-day coasts may therefore be due to a variety of geologic processes that operated on the land before sea level rose, and *not* to the marine processes operating today.

EROSION ALONG COASTS

Wave Refraction

Most waves in the ocean are generated by the wind in storms far offshore. As the waves move out from the storm area and approach the shore, they are commonly bent, or *refracted* (Figure 14.1). Wave refraction occurs along an irregular coast because, in shallow water, part of the wave begins to "drag bottom" and slow down, while the remainder of the wave in deeper water moves forward at normal velocity. As a result, wave refraction concentrates energy on headlands and disperses it in bays. One way to visualize wave refraction is to draw equally spaced rays (lines perpendicular to wave crests) and to note how the spacing changes as the waves are refracted. Energy will be concentrated in zones where the rays are close together and dispersed where the rays are far apart.

Longshore Drift

Longshore drift is generated as the waves strike the shore obliquely, and water and sediment, moved by the breakers, are transported up the beach in the direction of wave advancement.

When the energy is spent, water and sediment move as backwash down the slope of the beach, perpendicular to the shore. The sediment thus moves along the beach in a zigzag path (Figure 14.2).

The major landforms produced by wave erosion are wave-cut cliffs. Minor erosional features include sea stacks, sea arches, and sea caves. As a sea cliff recedes, a wave-cut platform develops.

DEPOSITION ALONG COASTS

Much of the sediment derived from erosion of the land is transported to the sea and deposited as deltas. It is then subject to the attack of waves and transported by longshore currents to form a beach. As longshore currents enter deeper water and lose energy, sand is deposited in the form of a *spit*. The spit may grow across the bay to form a barrier *bar* or barrier island. Between the barrier island and the mainland are the quiet waters of a lagoon. A break or passageway through the barrier island is a *tidal inlet*. Sediment moved through the tidal inlet will commonly be deposited as a *tidal delta*.

The major types of landforms developed along coasts are shown in Figure 14.3.

EVOLUTION OF SHORELINES

As a result of wave erosion and deposition, the configuration of a coast evolves until energy is distributed evenly along the coast so that neither large-scale erosion nor deposition is occurring. A coast like this is known as a *shoreline of equilibrium* and is characterized by a smooth, gently curving beach. Few present-day coasts approach this state of equilibrium, however, because most of the world's shorelines have been affected by the rise in sea level associated with the melting of the glaciers. Virtually all coastal areas have been submerged, and river valleys have become embayed or drowned.

CLASSIFICATION OF COASTS

A meaningful classification of coasts is based on the geologic processes that are responsible for the configuration of the shoreline. Two principal subdivisions are (1) *primary coasts,* formed by terrestrial processes, and (2) *secondary coasts,* formed by marine processes. Various subdivisions of these are shown in Figures 14.4 and 14.5.

FIGURE 14.1

Wave Refraction concentrates energy on headlands and disperses it across bays. Each segment of the unrefracted wave, AB, BC, and CD, has the same amount of energy. As the wave approaches shore, segment BC encounters the sea floor sooner than AB or CD and moves more slowly. This difference in the velocities of the three segments causes the wave to bend, so that the energy contained in segment BC is concentrated on the headland, while an equal amount of energy contained in AB and CD is dispersed along the beach.

FIGURE 14.2

Longshore Drift occurs where waves strike the beach at an oblique angle. As a wave breaks on the shore, sediment lifted by the surf is moved diagonally up the beach slope. The backwash then carries the particles back down the beach at a right angle to the shoreline. This action is repeated by each successive wave and transports the sediment along the coast in a zigzag pattern. Particles also are moved underwater in the breaker surf zone by this action.

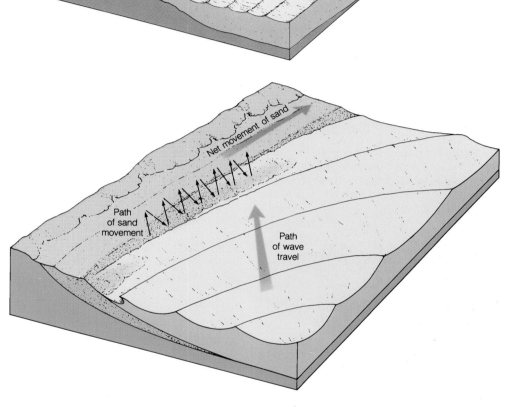

FIGURE 14.3

Major Types of Landforms Developed Along Coasts result from both erosion and deposition. The major erosional features are wave-cut cliffs and associated caves, sea stacks, and arches. Depositional features include beaches, spits, barrier bars, and organic reefs.

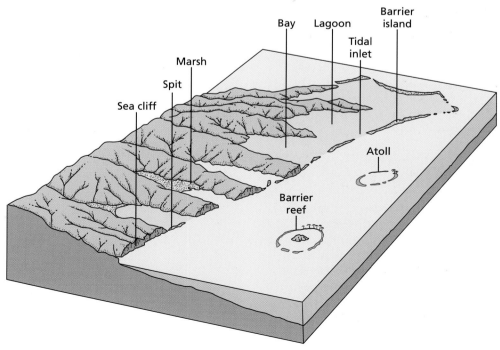

A. Stream Erosion: The configuration of the landscape was developed by stream erosion when the area was not covered by the sea. A subsequent rise in sea level, resulting from the melting of Pleistocene glaciers, drowned the river valleys.

B. Glacial Erosion: When glaciers melt and sea level rises, long arms of the sea extend many miles up the deep U-shaped valleys. These drowned glacial valleys, called fiords, are not greatly modified by wave action.

C. Stream Deposition: Where a major river enters the sea, it commonly deposits more sediment than waves and currents can carry away, so that new coastal land is added in the form of a delta.

FIGURE 14.4
Primary Coasts are configured by terrestrial processes.

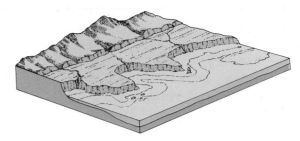

A. Wave Erosion: In areas where the rock is relatively uniform, wave erosion forms straight sea cliffs. If different rock types are present, wave erosion produces bays in the softer material and leaves the resistant rock projecting into the sea as rocky points.

B. Marine Deposition: Coasts built by sediment deposited by waves and currents are readily recognizable by beaches, barrier islands, spits, and bars.

C. Organically Built Coasts: The configuration of some shorelines is controlled by the growth of organisms such as coral reefs and mangrove trees.

FIGURE 14.5
Secondary Coasts are configured by marine processes.

PROBLEMS

As you answer the following questions, keep in mind the fundamental processes of wave refraction and long-shore drift, and remember that all shorelines change in such a way as to establish equilibrium with energy expended upon them.

0   0.5 miles

1. Coast—Central California

a. Study this area and label all of the coastal landforms you can identify.

b. Assuming that the present wave pattern is normal, show with arrows the prevailing pattern of longshore drift.

c. Note the jetties (long, narrow seawalls) constructed from head-land and positioned up along the main beach. How do these structures affect longshore drift?

d. Show how the energy between points A, B, C, and D will be distributed along the coast as a result of wave refraction. (A good approach to solving this problem is to draw lines along the crests of the major waves to emphasize their regional pattern, and then draw rays from points A, B, C, and D toward the shore perpendicular to the wave crests.)

e. Show the position of the ancient shoreline that existed when the wide wave-cut platform (on which the airstrip was built) was formed.

2. Newfoundland Coast—Canada

a. Study the coast and label the following features:

    (1) Headlands

    (2) Sea stacks

    (3) Wave-cut terrace

    (4) Tidal flat

    (5) Tidal channels

    (6) Sea cliff

b. Sketch a map showing how this coast will appear when it more closely approaches a state of equilibrium.

c. Erosion along a coast produces a variety of landforms that are soon modified or destroyed. Color the erosional feature that you believe will be the end product of wave erosion on this coast.

d. List all of the landforms that you recognize as the product of tides.

(Department of Energy, Mines, and Resources, Ottawa, Canada)

0   0.5   1 mile

3. Coastal Features—Quebec, Canada

a. Label the following features and explain their origin: (a) barrier bar, (b) tidal inlet, (c) tidal delta, (d) tidal channels, (e) lagoon, (f) tidal marsh, (g) bay-head delta, and (h) wave-cut platform.

b. What is the source of the sediment that is being deposited both seaward and landward of the inlet between the spits?

c. How will this coast change with time?

d. Compare this area with the Newfoundland coast. Which area represents the more advanced stage of coastal development?

(Department of Energy, Mines, and Resources, Ottawa, Canada)

0   0.5   1 mile

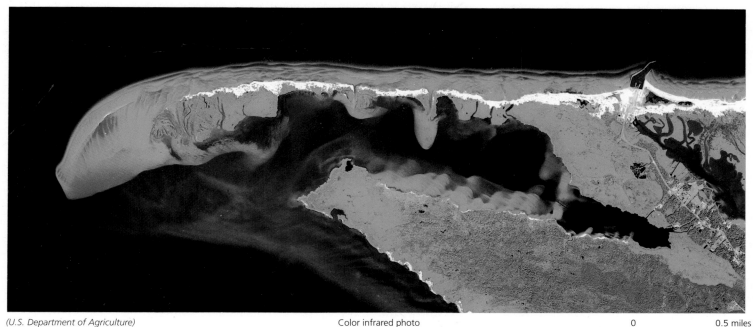

(U.S. Department of Agriculture)   Color infrared photo   0 _____ 0.5 miles

4. Spit—North Carolina

a. Show the orientation of the dominant wave pattern along this coast.

b. What evidence suggests that longshore drift occurs under water, as well as on the beach?

c. Note that the beach is broken by numerous inlets connecting the bay with the open ocean. What processes are probably responsible for these features?

d. Assuming that present processes continue, show by a series of lines how this coast will evolve.

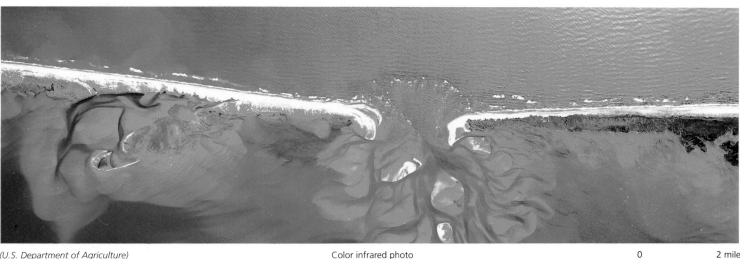

(U.S. Department of Agriculture)   Color infrared photo   0 _____ 2 miles

5. Coastal Features—North Carolina

a. The major landforms in the area are the result of longshore currents, which move sediment along the coast, and of tidal currents, which move sediment in and out of the tidal inlet. Study the shape of the various coastal features and show with a series of arrows the direction of sediment transport throughout the entire area.

b. Why is the tidal delta larger in the lagoon than in the open ocean?

c. Explain the origin of the island and associated sediment behind the barrier shown in the left-hand part of this photograph.

(NASA - Goddard Space Flight Center)   Landsat Mosaic   0   25   50 miles

6. Outer Banks— North Carolina

a. Label the following coastal features: (a) barrier bars, (b) spits, (c) tidal inlets, (d) lagoons, and (e) swamps.

b. This image was acquired Sept. 23, 1999, and shows the after-effects of flooding caused by Hurricane Floyd. Note the high volume of sediment carried by the floodwaters into the rivers, bays, and then into the open ocean. Show the direction of sediment movement throughout the area with a series of arrows. How does sediment get to the open ocean?

c. What processes transport the sediment in the lagoon?

d. What processes transport the sediment near the beach?

e. What processes transport the sediment beyond the beach?

(Space Shots Inc.)

Enhanced landsat mosaic

0   25 miles

7. Chesapeake Bay

a. Explain the origin of the coastline shown in this image.

b. What causes the water in some areas to appear light blue, whereas in others it appears to be dark blue?

c. What is the origin of the long, narrow bays? How will they be modified with time?

d. Make a series of sketches showing how the Chesapeake Bay area will appear in the future, as marine processes and sedimentation in the estuary continue.

e. How will the coast facing the open ocean change with time?

A. 7-13-77

(Courtesy Kelsey-Kennard Photographers Inc., Chatham, MA)

8. Beach Erosion—Chatham, Massachusetts (views looking north)

The town of Chatham, Massachusetts, is located at the southern end of Cape Cod. It is separated from the open ocean by a barrier island that terminates to the south as a spit. A break through the barrier occurred during a major storm on January 7, 1987. These photographs document the changes that occurred before the storm to May 1991.

a. Note the sand waves migrating at the end of the spit just below water level in (A) and (B). What is the immediate source of this sand?

b. What major changes occurred between the period of the initial break (B) 1988 and (C) 1991?

c. What is represented by the line of white waves seaward from the break in the barrier island shown in the photograph taken in 1988 (B)?

d. Note the buildings (homes and summer cottages) along the shore behind the barrier island (the arrow in A). How did this part of the shoreline change between 1988 and 1991?

e. What major changes occurred on the barrier island between 1988 and 1991? What caused those changes?

B. 5-9-88

C. 5-19-91

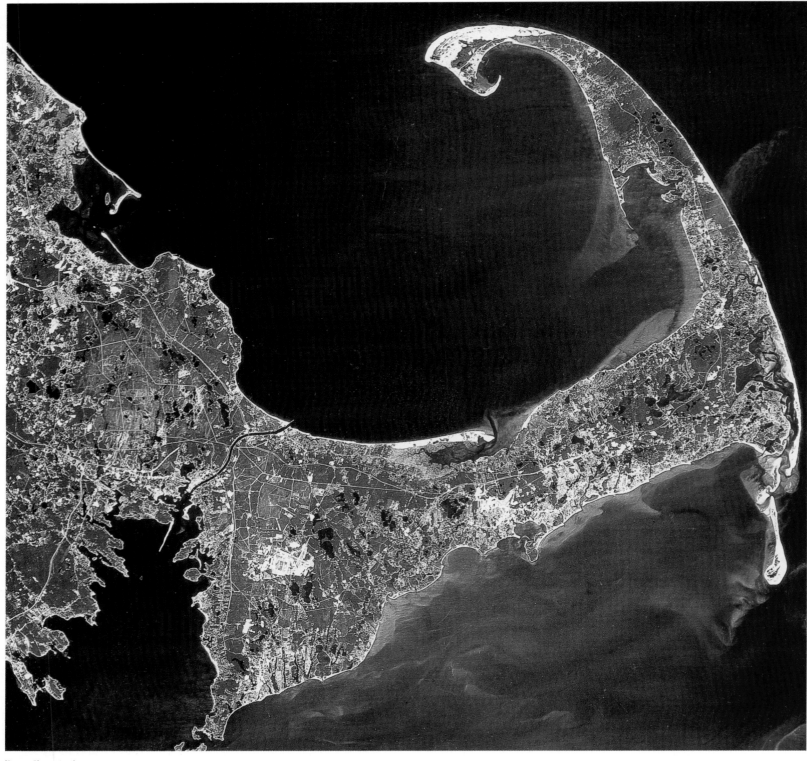

(Space Shots, Inc.)

9. Cape Cod

a. In which prevailing direction do the waves strike the coast on the northeastern part of this area?

b. Explain the origin of the curved spit along the northern tip of Cape Cod.

c. Study the small spits in this area and explain why they grow in various directions.

d. What evidence suggests that there has been a recent rise in sea level in this area?

10. Isles Dernieres Barrier System

Isle Dernieres is a barrier island system formed by the erosion of a previously active distributary system of the Mississippi River. (See page 108.) It is located about 100 km west of the present mouth of the Mississippi. It was once more than 36 km long, but storm surges and tidal inlet development have fragmented the barrier island into an arc of six small islands.

This series of maps are like snapshots in time, showing the configuration of the islands as they changed from 1853 to 1988. The oldest map (1853) represents a Coast and Geodetic Survey at a scale of 1:200,000. Subsequent mapping was more detailed, and the most recent surveys utilized aerial photographs.

Coastal erosion is continuous, but it is not a steady process. Bursts of erosion occur during and after the passage of major cold fronts, tropical storms, and hurricanes. Field measurements document that 20–30 m of coastal erosion have occurred during a single 3- to 4-day storm. These major storms may cause an overwash of beach sand into the lagoons, major breaks in the barrier bar, migration of beach dunes, and deposition of the sediment in new areas.

a. What major changes occurred between 1853 and the 1890's? What processes were responsible for those changes?

b. What changes occurred between the 1890's and 1934? What processes were responsible for those changes?

c. How did the area change between 1934 and 1988? What processes dominated during that time?

d. Estimate the amount of land loss in the interval between 1853 and 1988. What was the average annual rate of erosion during that time?

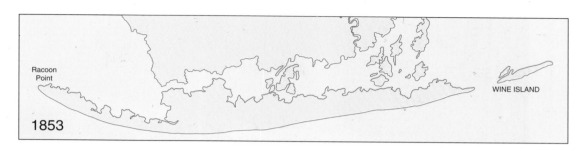

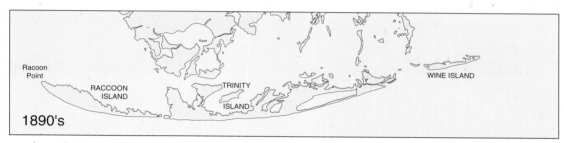

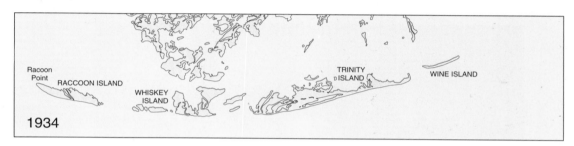

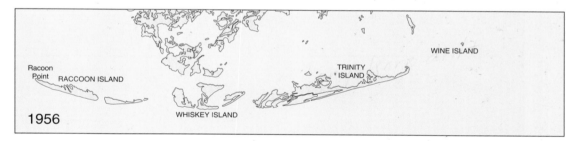

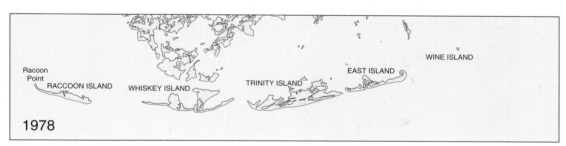

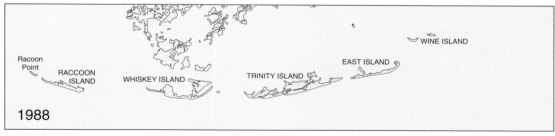

(U.S. Geological Survey)

(NOAA)

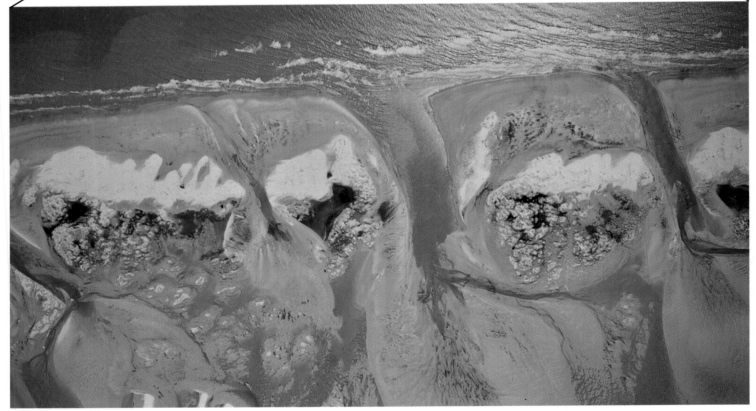

(NOAA)

11. Effects of Hurricanes— Texas

The high-altitude photograph at the top shows a barrier island off the coast of Texas. The low-altitude photograph shows a segment of the same island shortly after Hurricane Allen struck the coast on August 8, 1980. In addition to high winds, a hurricane is often accompanied by waves more than 20 ft higher than normal.

a. How did the hurricane modify the barrier island?

b. How did it modify the lagoon?

c. Is there any evidence for water (and sediment) moving back through the deep channels to the sea?

d. What evidence of abnormal wind activity can you find on the beach?

e. How will this area change with a return to normal shoreline processes?

f. What evidence of the hurricane will likely remain in the record as a body of rock?

OBJECTIVE

To recognize the major landforms produced by eolian activity and to understand the processes that control their development.

MAIN CONCEPT

Wind is capable of transporting large quantities of sand and dust which, when deposited, form distinctive types of dunes, or loess plains.

SUPPORTING IDEAS

1. The variety of dune types results from variations in sand supply, wind velocity, and wind direction and from the different characteristics of the surface over which the sand moves.
2. Dunes may be stabilized by the growth of vegetation.
3. The wind does not produce large-scale erosional features in most desert regions, but it can form deflation basins, which are significant landforms in some areas.
4. Large amounts of dust are transported high in the atmosphere and ultimately are deposited in the oceans or on land as loess.

WIND EROSION

The sand-blasting action of wind is capable of producing small-scale erosional features such as ventifacts, pits, and grooves, but wind action is not a powerful agent of erosion. The major form of wind erosion is *deflation*—the lifting and transporting of loose sand and dust. The dust is diffused upward into the atmosphere, and the sand is transported near the surface where it commonly accumulates as sand dunes. Deflation often results in the excavation of shallow depressions known as *deflation basins,* or *blowouts.* Deflation basins are roughly circular or elliptical and may grow to more than a mile in diameter and 50 ft deep. In addition, in some deserts, wind erosion may produce distinctive linear ridges called yardangs.

SAND DUNES

Although wind is relatively insignificant as an erosional agent, it is capable of transporting huge volumes of sand and is responsible for the formation of the great "sand seas" of the world's major deserts. Sand dunes can assume a variety of fascinating shapes and patterns, depending on such factors as sand sup-

ply, wind velocity, variability of wind direction, and the nature of the surface over which the sand moves (Figure 15.1). The most common varieties are:

1. Transverse dunes develop where there is a large supply of sand and a constant wind direction. These dunes may cover large areas and form "sand seas," so called because the wave-like dunes resemble a stormy sea.

2. Barchan dunes form where the supply of sand is limited and the wind direction is constant. They are crescent-shaped with the tips, or horns, of the crescent pointing downwind.

3. Longitudinal dunes are long, parallel ridges of sand, elongate in the direction of the prevailing wind. They are caused by winds blowing from two slightly different directions.

4. Star dunes are mounds of sand having a high central point from which several ridges radiate. They form where winds blow from three or more directions.

5. Parabolic dunes typically develop along coastlines where vegetation partly covers a ridge of windblown sand derived from the beach. Where vegetation is sparse or absent, small deflation basins are produced by strong onshore winds. The depression grows larger as

more sand is removed. Usually, the sand accumulates downwind, forming a crescent-shaped ridge. In map view, parabolic dunes are similar to barchan dunes, except that the tips of a parabolic dune point upwind.

PROBLEMS

In this exercise, we will study aerial photographs and satellite images showing both erosional and depositional landforms produced by the wind. The satellite images are especially important because they show regional views of sand seas in the United States, plus the Sahara and the Namib deserts of Africa.

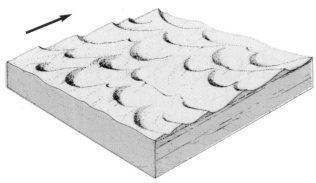

A. Tranverse Dunes

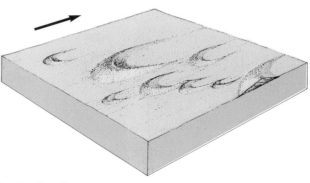

B. Barchan Dunes

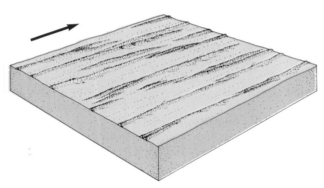

C. Longitudinal Dunes

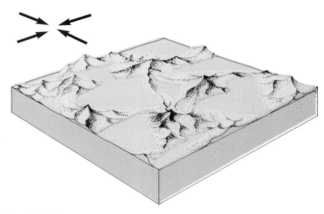

D. Star Dunes

(Arrows indicate
prevailing wind
direction)

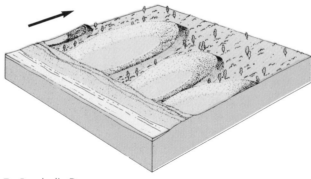

E. Parabolic Dunes

**FIGURE 15.1
Dune Types**

1. Deflation Basins—West Texas

This area is in the Great Plains province of the central United States. The rocks are mostly sandstones, siltstones, and shales, all of which are weakly cemented with calcium carbonate. The climate is semiarid.

a. What is the average size of the depressions?

b. How do they differ from sinks (pages 119–120) and kettles (page 137)?

c. How would rainwater entrapped in these basins influence their development?

d. What geologic evidence shown in the photograph would support the conclusion that the depressions are deflation basins and not sink-holes? (Note the drainage patterns shown in the southwest corner and the evidence of semiarid climate, as indicated by the vegetation.)

(U.S. Department of Agriculture)

0   1 mile

2. Sand Seas—North Africa

a. How many dune types can you recognize in this area?

b. Plot the prevailing wind direction with a series of arrows. What causes the desert conditions in North Africa?

c. How does this dune field compare with that of the sand hills in Nebraska (page 160)? Did they originate in the same way?

d. What evidence indicates that the landscape, now buried beneath the "sea of sand," was originally eroded by running water?

e. How has the topography of the mountainous landscape influenced the types of dunes and their distribution?

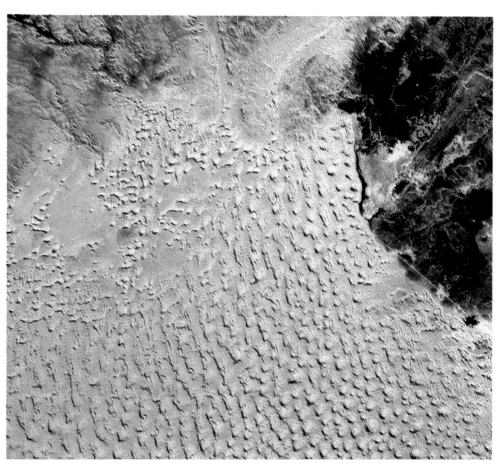

(U.S. Department of Agriculture)   Enhanced Landsat image   0   5 miles

3. Sand Dunes—California

a. What type of sand dune is common in this area?

b. What factors influence the formation of this type of dune?

c. If the supply of sand were greatly increased, what type of dune would most likely evolve in this area?

d. What is the direction of the prevailing wind?

e. What is the most probable source of sand in a desert area such as this: (a) wind abrasion of the general surface, (b) mass movement and slope retreat of local valley walls, or (c) sediment transported by a drainage system?

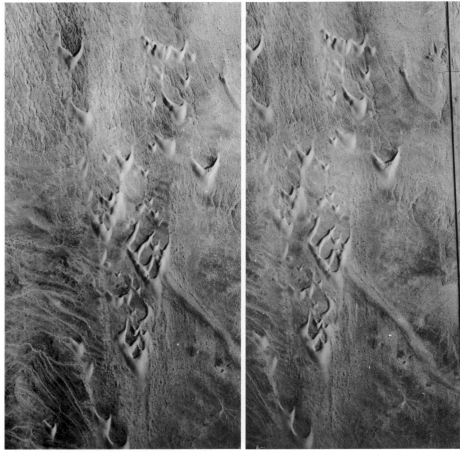

(U.S. Department of Agriculture)

0   1 mile

4. Coastal Dunes—Oregon

a. What type of sand dune is most common in this area?

b. Sketch a map showing the major dune fields according to their relative age. How many generations of dune systems can you identify?

c. Explain the origin of the elongate ridges and furrows just inland from the beach.

d. Sediment is being transported by three systems: eolian, beach, and stream. Show, with arrows, the direction of sediment transport in each of these systems.

e. Which system of sediment transport dominates in this area: fluvial or beach?

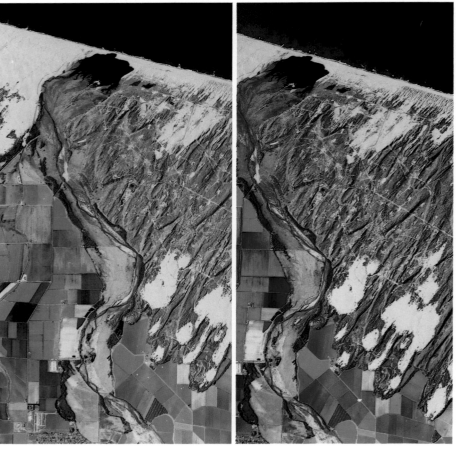

(U.S. Department of Agriculture)   Color infrared   0   0.5 miles

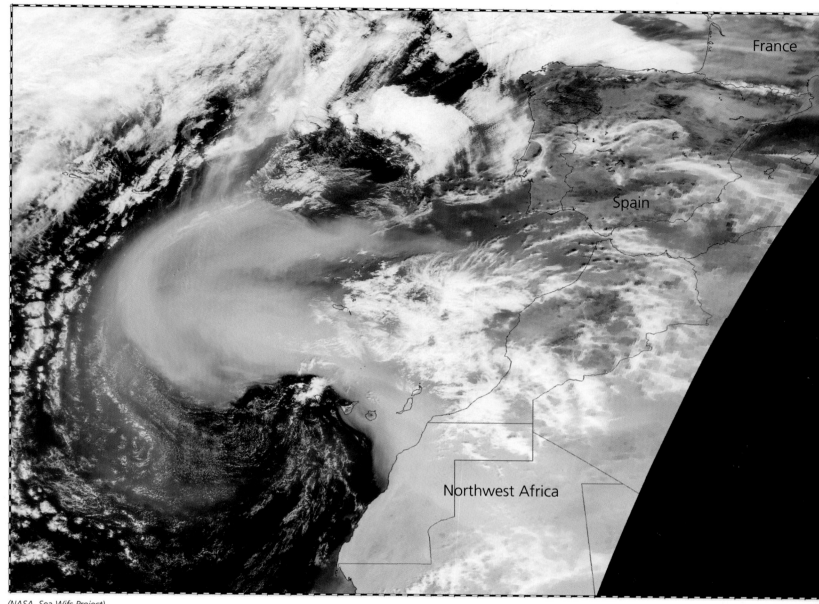

France

Spain

Northwest Africa

(NASA, Sea Wifs Project)

5. Dust Storm Over the Atlantic Ocean

This massive dust storm blowing off the Sahara Desert covers hundreds of thousands of square miles of the eastern Atlantic Ocean. The image was acquired February 26, 2000 when it reached over 1,000 miles out into the Atlantic. Such storms result from rising warm air that can lift dust as much as 15,000 ft into the atmosphere. These dust storms commonly reach as far as the Caribbean and are believed to cause a decline in coral reef growth, and may even play a role in the frequency and intensity of hurricanes.

a. Study the patterns of global atmospheric circulation from the figures in your text. Why is northern Africa a desert?

b. Why do dust storms in the Sahara blow to the west?

c. What determines the distinctive shape of this storm?

d. If this type of storm moved over a continent, what type of deposit would be produced when the dust settled?

e. What type of deposit will be produced when the dust settles to the bottom of the Atlantic?

159

6. Sand Hills—Nebraska

The Sand Hills of Nebraska were formed during the last ice age, when wind and sand supply were greater, and precipitation was less. In most areas, the dunes are now stabilized by the growth of prairie grass. The area outlined here is shown at a larger scale below.

a. Draw a series of arrows to indicate the wind direction during dune development. Was the wind direction constant or variable?

b. How have the dunes affected the drainage in this area?

c. Are the dunes older or younger than the North Platte River (the large stream near the bottom of the image)?

d. Are the dunes older or younger than the small streams shown near the eastern border of the image?

e. How will this area change in the future?

(Earth Satellite Corp)   Landsat image   0   20 miles

7. Sand Hills—Nebraska

a. This is a high-altitude, infrared photograph of the area outlined on the Landsat image above. Study this photograph and show, with a series of arrows, the wind direction in this area when the dunes were developing.

b. Identify the type of dune that occurs in this area.

c. Explain the origin of the lakes.

d. How have the geologic processes that developed this topography been modified in the last several thousand years?

e. How will this area change in the future?

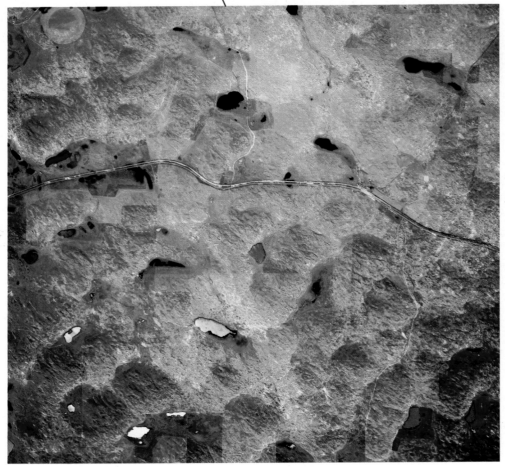

(U.S. Department of Agriculture)   Color infrared photo   0   1 mile

0   5 miles

8. Sand Dunes of the Namib Desert—Namibia

This is the northeastern part of the Namib Desert in southwestern Africa. The dune field is bounded on the north by the Kuiseb River and on the east by the Great Escarpment. Two major dune types occur in this area.

a. Outline, on the photograph, the boundaries of the major dune types. What visual evidence suggests that longitudinal dunes break up into star dunes here?

b. Is the prevailing wind constant or variable?

c. Why does the dune field terminate abruptly at the Kuiseb River?

d. What is the origin of the main trunk stream of the Kuiseb River?

161

OBJECTIVE

To become acquainted with the major structural features of Earth's crust and with the outcrop patterns they produce, and to understand what each major feature implies about crustal mobility.

MAIN CONCEPT

Each major geologic structure—anticline, syncline, dome, basin, or fault—has a specific geometric form. When the structure is exposed at the surface, it produces distinctive outcrop patterns that can be recognized on an aerial photo-graph or geologic map. These patterns provide a basis for interpreting an area's geologic history.

SUPPORTING IDEAS

1. Geologic maps show the distribution of rock formations, faults, and other structural features as they appear exposed at the surface. Each map is a scale model of the rock bodies of the crust and a fundamental tool for analyzing and interpreting geologic data.
2. Many important aspects of a region's geologic history can be interpreted from geologic maps, aerial photographs, and remote sensing data.

GEOLOGIC MAPS

An accurate idea of the size, shape, and extent of rock bodies is essential if we are to interpret the geologic history of a region. Units of sandstone, limestone, or shale, which are known as *formations,* may range in thickness from a few feet to several thousand feet and may cover an area of more than 200,000 mi². Formations, therefore, exist as extensive layers of rock that can cover several states.

The problem of studying the distribution and structure of rock bodies is similar to the problem of studying land-forms; it is basically a problem of scale. An understanding of the configuration and regional extent of an entire rock body can be obtained only by careful geologic mapping. Geologic maps

show the distribution of rock units as they crop out at the surface. Each rock unit is shown by a specific color or by a tone of gray. A legend accompanies the map to define the symbols and col-orations and to indicate the age and stratigraphic relationships of the rock units. The conventional geologic map symbols for common structural features are shown in Figure 16.1.

To understand the surface expres-sion of these various features, you should first review the basic three-dimensional geometry of the major structures (Figure 16.2) and the discus-sion in your textbook.

Map Patterns

The major types of structural features produce distinctive patterns on geologic

maps. Examples of various common structural patterns are shown in Figures 16.3–16.12. On each page in this sec-tion, the figure shows a block diagram consisting of a bird's eye view of the structural feature and its topographic expression. Above the diagram, the out-crop pattern of each rock unit shown on the block diagram is projected onto a horizontal plane to represent a geo-logic map in perspective. This arrange-ment makes it easy to relate the struc-ture shown on the block diagram to the map patterns. The careful study and understanding of these diagrams is an important prerequisite to the effective interpretation of geologic maps.

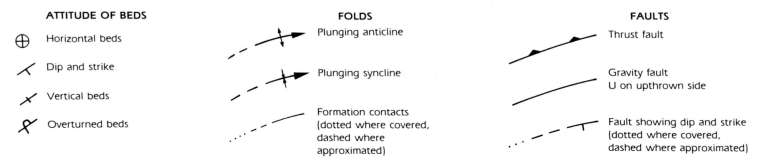

ATTITUDE OF BEDS	FOLDS	FAULTS
⊕ Horizontal beds	Plunging anticline	Thrust fault
⤨ Dip and strike	Plunging syncline	Gravity fault U on upthrown side
⤯ Vertical beds	Formation contacts (dotted where covered, dashed where approximated)	Fault showing dip and strike (dotted where covered, dashed where approximated)
⤰ Overturned beds		

FIGURE 16.1
Geologic Map Symbols

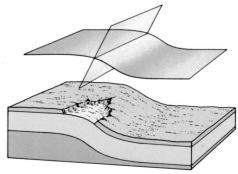

A. Monocline

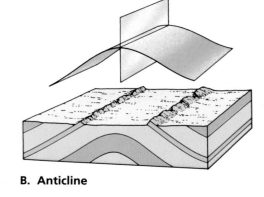

B. Anticline

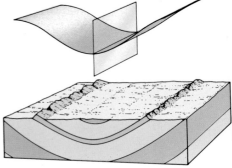

C. Syncline

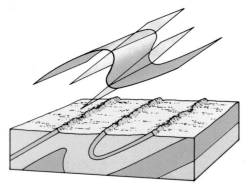

D. Overturned Anticline and Syncline

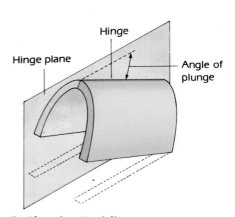

E. Plunging Anticline

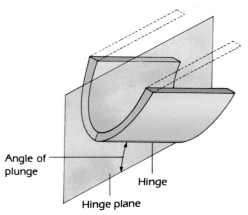

F. Plunging Syncline

FIGURE 16.2
Three-Dimensional Geometry of Folds

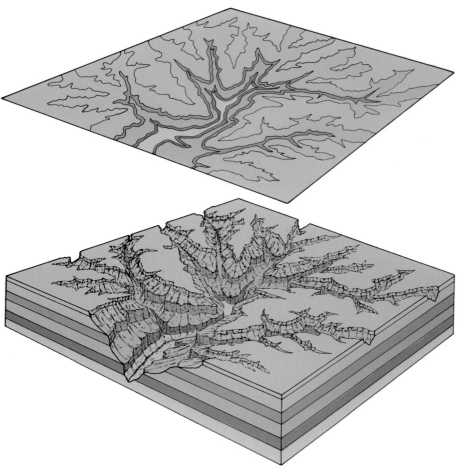

FIGURE 16.3
Outcrop Patterns of Horizontal Strata

A dendritic, or tree-like, drainage pattern characteristically develops on horizontal strata. As erosion proceeds, the river system cuts canyons, or valleys, in which a succession of older rock units is exposed. The basic outcrop patterns of horizontal strata exposed on the valley walls, therefore, produce a dendritic pattern on the geologic map. The contacts of the rock units of horizontal strata are essentially parallel to the topographic contour lines.

Cliffs develop on resistant formations, and slopes develop on the weaker, nonresistant rock layers. This produces variations in the width of the map outcrop patterns. On a steep cliff, the upper and lower contacts (as seen on the map) will appear to be close together, whereas on a gentle slope of the same formation, the contacts will appear to be farther apart. Note that the red, blue, and green layers in the block diagram are essentially the same thickness, but the red and dark green layers form cliffs, and their map patterns form thin bands compared to the wide bands formed by the blue and light-green layers which form slopes. As seen on a map, the width of an outcrop belt of horizontal strata therefore depends not only on the unit's thickness, but also on the topography.

Gently dipping strata will develop the same basic outcrop pattern as horizontal beds. However, the contacts between rock units in gently dipping strata, when traced far enough up a valley, cross topographic contours and form a large V-shaped pattern that points in the direction that the beds dip.

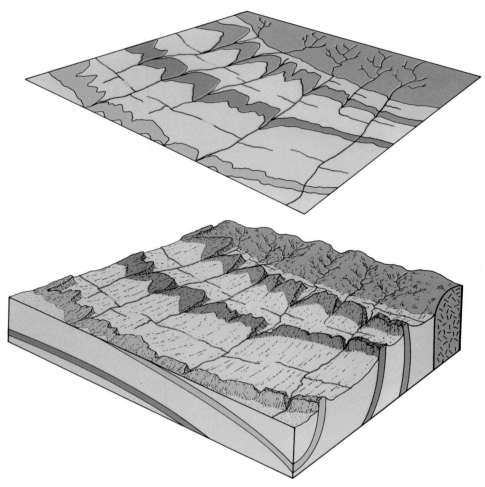

FIGURE 16.4
Outcrop Patterns of Inclined Strata

When a horizontal rock sequence becomes tilted and part of it is removed by erosion, the outcrop patterns of the rock units (viewed on a regional scale) become dramatically different. The rock units appear as roughly parallel bands. Important variations in the basic map pattern develop in areas dissected by erosion. These variations should be analyzed carefully, for they provide important information about the subsurface structure. Take a minute to follow each of the streams on the block diagram from its headwaters downstream. Note the stream patterns and how they are related to the rock formations.

When dipping strata are traced across a valley, a V-shaped outcrop pattern that points in the direction of the dip is produced as you can see on the block diagram. (Exceptions to this rule are possible if the degree at which the beds dip is less than the gradient of the valley, but such conditions are rarely

encountered.) The size of the V is inversely proportional to the magnitude of the dip, and we can make the following generalizations:

1. Low-angle dip produces a large V (left side of block diagram).

2. High-angle dip produces a small V.

3. Vertical dip produces no V (right side of block diagram).

Careful study of the figure will reveal more relationships basic to the geologic map patterns:

1. Older beds dip toward younger beds (unless the sequence is overturned).

2. Outcrop width depends on

    a. Thickness of the beds

    b. Dip of the beds (the lower the dip, the wider the outcrop)

    c. Slope of the topography (the steeper the slope, the narrower the outcrop)

All these relationships are shown on the block diagram and geologic map above. Note that the beds on the right side of the diagram are nearly vertical, but when traced across the diagram to the left margin, the dip decreases to about 15°. Now look carefully at the map, especially the blue and red formations. Where the dip is vertical, the width of the outcrop belt is narrow and there are no pronounced Vs where streams cut across the formations. As the dip decreases toward the left of the diagram, the width of the outcrop belt increases and the size of the Vs increases.

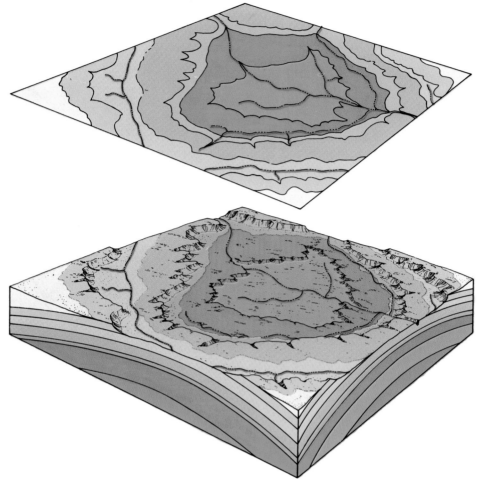

FIGURE 16.5
Outcrop Patterns of a Dome

Eroded structural domes form a roughly circular or elliptical outcrop pattern in which the beds dip away from the crest, or central part of the dome. Domes range in size from small upwarps that are a few feet in diameter, to regional features that cover hundreds or thousands of square miles. As you can see, the oldest beds are exposed in the central part of an eroded dome, and progressively younger formations are exposed outward from the center of the structure.

Streams cutting across the dome enable us to apply the rule of Vs to determine the direction of dip. As can be seen on the map, the Vs point away from the center of the dome. If the relative age of each rock unit is shown on the map, a dome is easily identifiable because the older rocks are in the center of the structure. Note also that a cliff or scarp formed on the beds of a dome faces inward, and the dip of the slope formed on the top of major units is inclined away from the center of the dome.

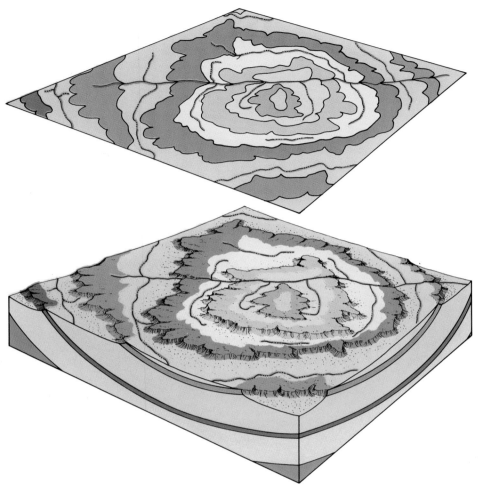

FIGURE 16.6
Outcrop Patterns of a Basin

A structural basin, when eroded and exposed at the surface, displays an elliptical or circular outcrop pattern similar in many respects to that of an eroded dome (Compare Figures 16.5 and 16.6.) But two major features on the map enable us to readily distinguish a basin from a dome:

1. *Younger* rocks crop out in the *center* of a *basin,* whereas older rocks are exposed in the center of a dome.

2. Vs in the outcrop pattern *point toward the center of a basin,* whereas they point away from the center of a dome.

In addition, the cliff or scarp formed on the resistant rocks of a basin faces outward, and the dip slope is inclined toward the center of the structure. This is exactly the opposite of the direction in which the slope is inclined in an eroded dome.

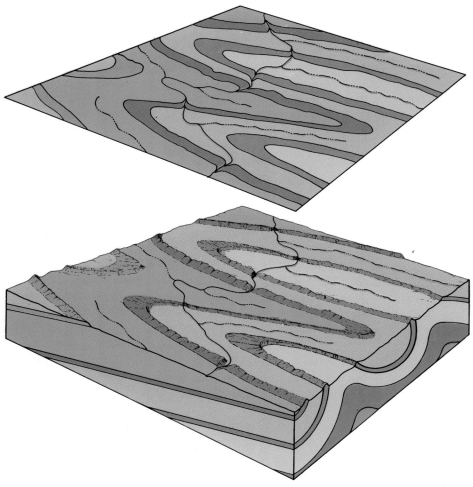

FIGURE 16.7
Outcrop Patterns of Plunging Folds

The outcrop pattern of a series of plunging folds forms a characteristic zigzag (Z-shaped) pattern. A plunging anticline forms a large V-shaped outcrop pattern with the apex (or nose) *pointing* in the direction of plunge. Also note that the oldest rocks in a plunging anticline are exposed in the center of the fold.

In contrast, a plunging syncline forms a large V-shaped outcrop pattern that *opens* in the direction of plunge, and the youngest rocks are in the center of the fold. Note also that where streams cut across the folds, the rule of Vs applies, and the direction in which the beds dip can be determined.

In the figure, the basic outcrop pattern of a plunging anticline is best shown in the outcrop pattern of the orange formation and the brown formation. Both form a V-shaped pattern, with the apex pointing in the direction of plunge. The basic outcrop pattern of a plunging syncline is best shown by the brown formation in the center of the diagram. This fold is situated between the adjacent plunging anticlines.

Note that the plunging anticlines and synclines are side by side, so that outcrops form a zigzag pattern. Also, study the drainage pattern. Does it reflect the underlying structure?

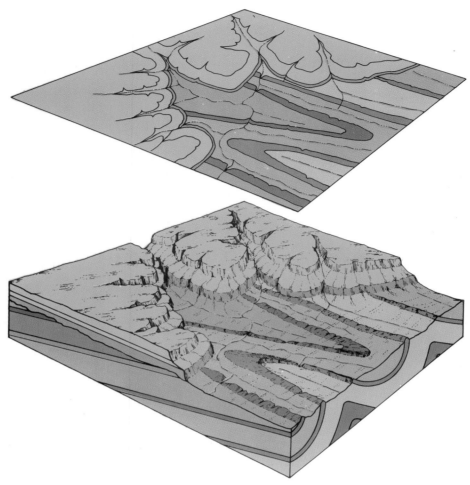

FIGURE 16.8
Outcrop Patterns of an Unconformity

Angular unconformities can be recognized on geologic maps by interruptions, or discontinuities, in the outcrop patterns. The older structures are partly covered by younger strata, so that, on the geologic map, the contacts of the older structures will terminate abruptly against the patterns of the overlying younger beds.

In the block diagram, the oldest sequence (shown in shades of blue, orange, and brown) has been warped into plunging folds, eroded, and subsequently covered by a younger sequence of strata (shown in shades of green and tan). A second period of erosion has partly removed the younger strata and exposed segments of the older folds. *The angular unconformity is located at the base of the sequence of younger horizontal strata.* All of the map patterns of the older strata terminate against this contact.

Vs in the trace of the unconformable surface indicate the direction in which the unconformity dips. In this example, the unconformity is nearly horizontal, so its outcrop pattern is dendritic. Note the change in the stream patterns formed on the younger rocks compared with the older rocks.

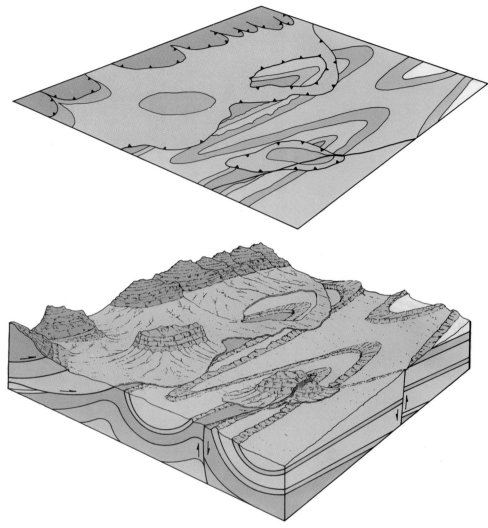

FIGURE 16.9
Outcrop Patterns of Normal and Thrust Faults

Fault patterns on geologic maps are distinctive in that they abruptly offset structures and terminate contacts between rock formations. They are expressed on a geologic map by *heavy* contact lines in order to be easily recognized.

Thrust faults generally dip at a low angle. Because of the low-angle dip, the pattern of the fault trace is characteristically irregular and similar in many respects to the trace produced by low-dipping angular unconformities (preceding page). In Figure 16.9, thrust faults are located at the base of the formations colored purple and blue. The trace of a thrust fault commonly forms a V across valleys, with the V pointing in the direction in which the fault dips.

Erosion may form windows (fensters) through the thrust sheet, exposing the older rocks, or may produce isolated remnants (klippen) in front of the main thrust block. Note that hachure marks are used to designate the overthrust block on the geologic map.

Normal and *reverse faults* usually dip at a high angle, so that their outcrop patterns are relatively straight but commonly form a zigzag pattern on a regional scale. Because older rocks are generally exposed on the upthrown block, the relative movement, on most high-angle faults, can be determined from the map relationships alone.

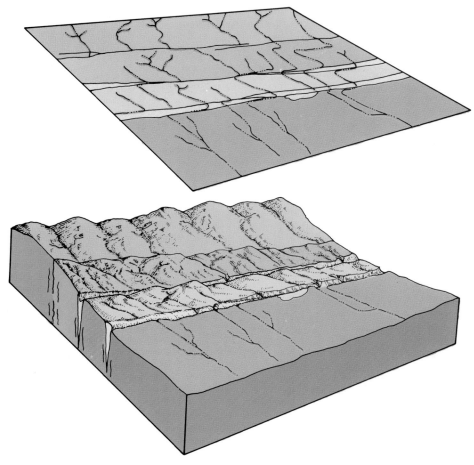

FIGURE 16.10
Outcrop Patterns of Strike-Slip Faults

In a *strike-slip fault,* the displacement is parallel to the strike of the fault plane. This type of movement is commonly associated with horizontal stress related to the separation of lithospheric plates. Displacement on strike-slip faults can reach several hundred miles, so that rock types of very different structure and geologic characteristics may be placed side by side. The trend of strike-slip faults is typically straight, in contrast to the irregular trace of thrust faults and the zigzag trace of normal faults. Small slivers or slices of foreign rock bodies may be caught in the strike-slip fault zone and are commonly expressed either as elongate troughs or ridges, depending on the nature of the rock mass.

The lateral displacement of the crust in strike-slip faults generally does not produce high scarps. The fault line is, however, commonly marked by structural and topographic discontinuities, linear ridges and valleys, and offset drainage patterns. The offset drainage is usually very significant because it indicates the direction of displacement.

Study the offset drainage pattern on the geologic map. In which direction did each block move relative to its neighbors?

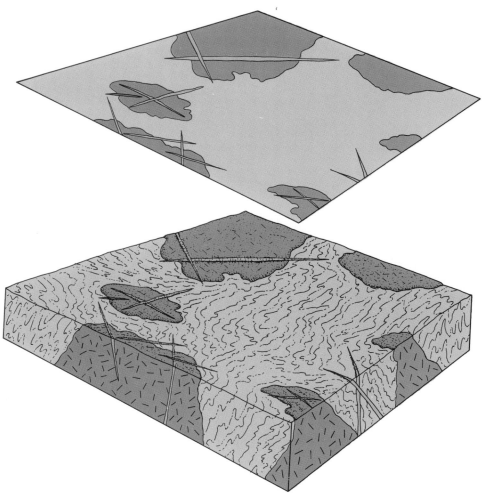

FIGURE 16.11
Outcrop Patterns of Intrusive Rocks

Intrusive igneous bodies may be either concordant or discordant with the surrounding rocks. The larger intrusives, such as stocks and batholiths, are characteristically discordant. Their outcrop patterns appear on geologic maps as elliptical or roughly circular areas that cut across the contacts of the surrounding formations. Smaller discordant intrusives, such as dikes, are characteristically tabular. Dikes generally follow joint or fault systems and appear on geologic maps as straight, narrow bands that cut across adjacent rock bodies. Some dikes, however, are lens-shaped and appear so on the map. Concordant igneous bodies, such as sills and laccoliths (not shown), have contacts parallel to those of the surrounding rock formations.

The relative age of igneous bodies can be recognized, on a geologic map, from the crosscutting contacts: the younger intrusives cut the older ones.

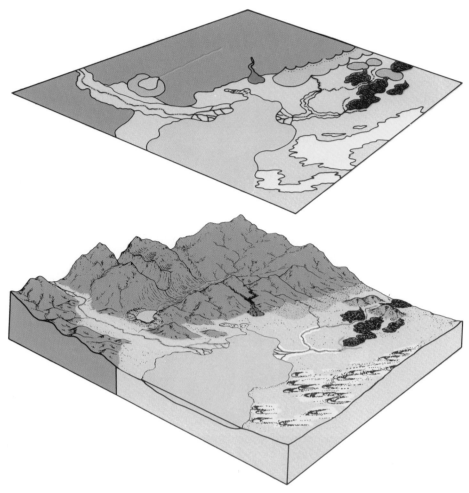

FIGURE 16.12
Outcrop Patterns of Surficial Deposits

Surficial deposits are various types of sediment or volcanic material that form on the surface of the present landscape. They are characteristically thin and rest upon the underlying bedrock. The principal types of surficial deposits are stream sediments, windblown sand and loess, landslide debris, glacial sediments, beach sand and other shoreline sediments, and recent volcanic deposits.

Surficial deposits may or may not be shown on a geologic map, depending on the objectives of those who made the map. Some geologic maps do, however, emphasize surficial deposits when the deposits are important to engineering projects or have significant environmental implications.

On a geologic map, surficial deposits typically form an elongate, irregular pattern and are commonly associated with the major processes actively operating on the landscape or that were active in the recent geologic past. In the diagram above, note the relationship between glacial deposits and the glacial valley, between stream deposits and the present stream valleys, and between landslide deposits and the mountain front. Surficial deposits rest unconformably on the underlying bedrock and do not extend very far into the subsurface.

PROBLEMS

1. Study the maps and block diagrams in Figures 16.3–16.9 and complete the geologic maps by adding the appropriate symbols to express geologic structure.

2. Estimate the angle of dip where the beds have been deformed and plot appropriate dip and strike symbols to express further the geologic structure.

3. On Figure 16.5, note that the blue formation is nonresistant and forms a basin or valley in the central part of the dome. Draw a cross section across the dome, showing how the surface of the land would appear if the blue formation were resistant and the red formation were nonresistant.

4. On Figure 16.6, label the formations shown in the figure according to relative age, with 1 = oldest, and so on.

5. On Figure 16.8, what evidence, on the map alone, indicates that:

    a. the green formations are younger than the blue and orange formations?

    b. the green formations are horizontal?

    c. the blue and orange formations are folded?

6. On Figure 16.9, show the direction of relative movement of the fault blocks (U = up, D = down).

7. Refer to Figure 16.10.

    a. What map evidence indicates the direction of displacement on the faults?

    b. What evidence indicates that the faults are strike-slip and not thrusts?

8. Refer to Figure 16.11.

    a. How do you know that the large intrusive bodies are younger than the metamorphic rocks?

    b. What is the youngest rock body? How did you determine your answer?

9. On Figure 16.12, label the following types of surficial deposits: (1) stream sediments, (2) landslide debris, (3) windblown sand, (4) alluvial fans, (5) lava flows, (6) volcanic cones, and sediment filling a structural basin.

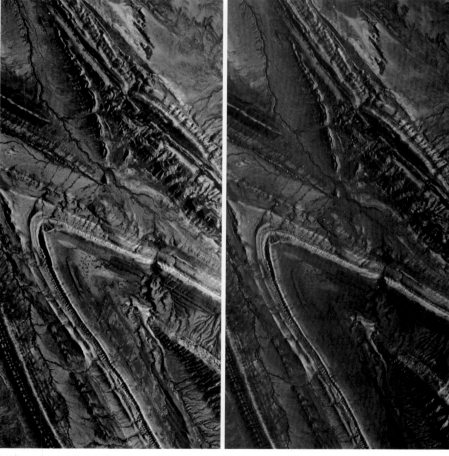

(U.S. Department of Agriculture)

0   0.5 miles

10. Geologic Structures— Wyoming

This area illustrates the fine details produced by erosion on folded rocks in an arid region. Resistant formations form ridges, nonresistant units form valleys, and specific textures caused by stream patterns form on the different rock units.

    a. Draw a geologic map of the area, by tracing the contacts of the major resistant formations. Show dip and strike symbols and the hinge line of all folds.

    b. Draw a geologic cross section through the structure, from the lower left corner to the middle of the photograph.

    c. Trace the major drainage pattern with a blue pencil. How has the structure controlled the drainage?

11. Geologic Structures—Kansas

a. Draw a geologic map of the area shown in stereoscopic view by tracing the contacts between several rock bodies, which are expressed by the selective growth of vegetation.

b. Number the beds according to relative age, with 1 as the oldest.

c. What is the structure of the bedrock?

d. What surficial deposits occur in this area?

e. Sketch a geologic cross section across the area.

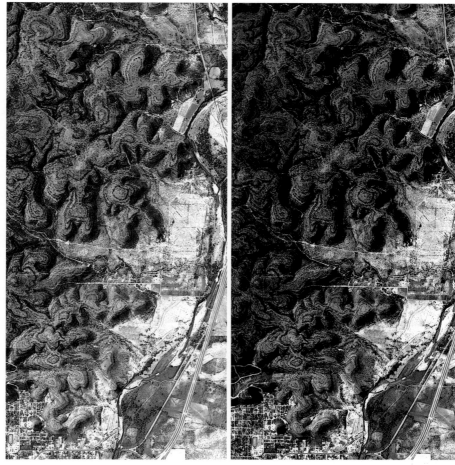

(U.S. Department of Agriculture)

0   0.5 miles

12. Geologic Structures—Montana

Two major rock sequences can be recognized in this area: (1) an older sequence that is folded and eroded into ridges and valleys, and (2) a younger, horizontal sequence.

a. Trace the contact between these major rock sequences.

b. What is this contact called?

c. Draw a geologic map of the area and outline the sequence of major geologic events.

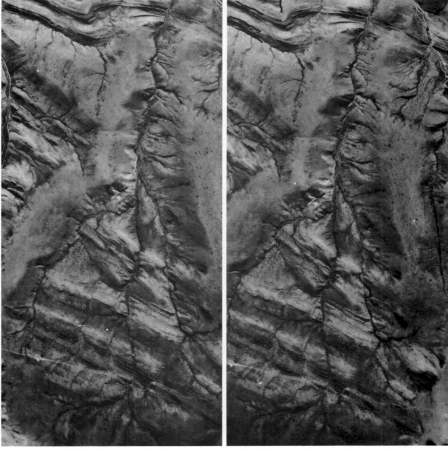

(U.S. Department of Agriculture)

0   1 mile

13. Geologic Structures—Utah

a. Trace the major faults shown in this area. Are they thrust, normal, or strike-slip faults?

b. Draw a generalized cross section across the fault system. What is the origin of the valleys?

c. What type of stress (compression, tension, or shear) most likely produced these faults?

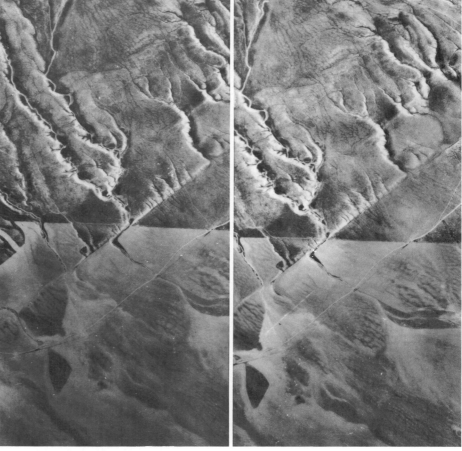

(U.S. Department of Agriculture)

0   0.5 miles

14. Geologic Structures— California

a. What is the major structural feature shown in this area?

b. Trace the fault lines with a red pencil and the drainage lines in blue. Is the fault in a single plane or in a zone of several parallel planes?

c. What type of fault occurs in this area?

d. Study the stream pattern on both sides of the faults. How does the stream pattern show the relative movement along the faults?

e. Draw arrows to show the relative movement on either side of the fault plane and determine the amount of recent displacement.

(U.S. Department of Agriculture)

0   0.4 miles

15. Geologic Structures— Canada

Three major rock bodies are exposed in this area: (1) metamorphosed sediments (dark gray), (2) granite (light tones), and (3) dikes (long, narrow bands of dark-gray tones).

a. Draw a geologic map of the area shown in stereoscopic view by tracing contacts of the major formation on the photo. What types of intrusive bodies are formed by the granite?

b. Study the contacts between the granite and the dikes, and determine the relative age of these rock bodies.

c. Which rock types are most resistant to weathering? Which weather most rapidly?

d. Draw an idealized cross section across the area, from the lower left corner to the upper right, to show how the rock bodies might appear below the surface.

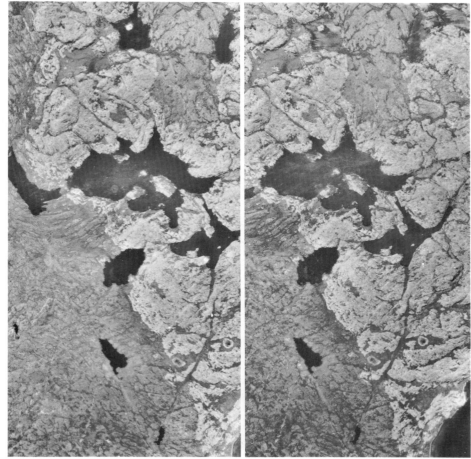

(Department of Energy, Mines and Resources, Ottawa, Canada)

0   1 mile

16. Geologic Structures— Wyoming

a. Draw a geologic map of this area by drawing the contacts between rock bodies expressed by resistant ridges or low valleys.

b. Estimate the angle of the dips, and plot several dip and strike symbols on your map.

c. Draw a north–south geologic cross section through the structure.

d. What is the structure of the rocks that form the flat surface surrounding the circular hill?

e. Show how the cross section would change if the oldest formation were weak and nonresistant, and the younger units were resistant to erosion.

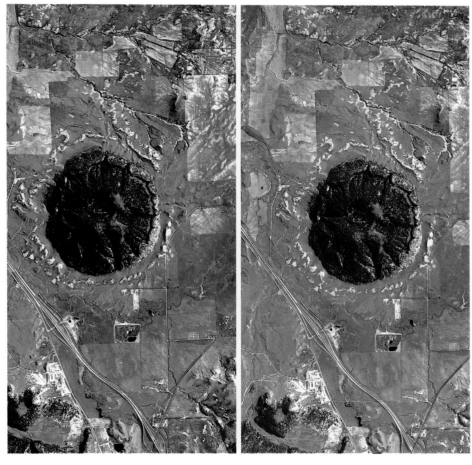

(U.S. Department of Agriculture)

0   0.5 miles

17. Geologic Map of Part of the Grand Canyon—Arizona

This is a section of the geologic map of the Grand Canyon. It is in the vicinity of the Visitor's Center on the south rim, which is the most popular part of the canyon. Study the general patterns of rock formations shown in various colors and refer again to Figure 16.3.

a. How do the outcrop patterns indicate that most of the Paleozoic strata in the area are horizontal?

b. What is the approximate thickness of the total sequence of Paleozoic rocks in the area? How did you arrive at your answer?

c. Draw a geologic cross section along line A–A'.

d. Note the relationship of rock type to cliff and slope topography on the map and on your geologic cross section. Which rock types develop cliffs and which develop slopes?

e. Note the outcrop pattern of the Temple Butte Limestone. Why is this pattern discontinuous?

f. How many unconformities are shown on the map? Between which units are they located?

g. What type of faulting is shown on the map?

h. What effect has faulting had on the course of the Colorado River and on the development of tributary valleys?

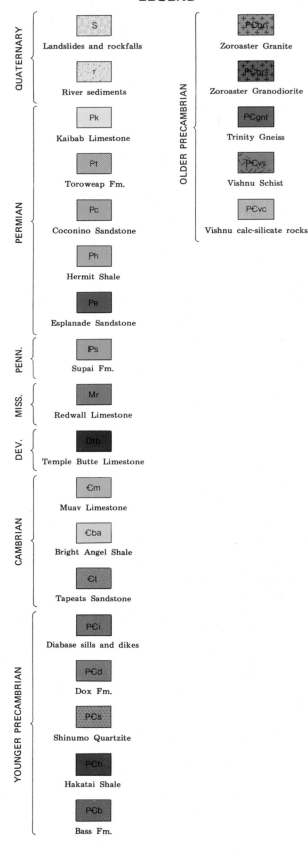

LEGEND

QUATERNARY
- S — Landslides and rockfalls
- r — River sediments

PERMIAN
- Pk — Kaibab Limestone
- Pt — Toroweap Fm.
- Pc — Coconino Sandstone
- Ph — Hermit Shale
- Pe — Esplanade Sandstone

PENN.
- IPs — Supai Fm.

MISS.
- Mr — Redwall Limestone

DEV.
- Dtb — Temple Butte Limestone

CAMBRIAN
- €m — Muav Limestone
- €ba — Bright Angel Shale
- €t — Tapeats Sandstone

YOUNGER PRECAMBRIAN
- PC̵i — Diabase sills and dikes
- PC̵d — Dox Fm.
- PC̵s — Shinumo Quartzite
- PC̵h — Hakatai Shale
- PC̵b — Bass Fm.

OLDER PRECAMBRIAN
- PC̵gr — Zoroaster Granite
- PC̵grd — Zoroaster Granodiorite
- PC̵gnt — Trinity Gneiss
- PC̵vs — Vishnu Schist
- PC̵vc — Vishnu calc-silicate rocks

CI 80 ft   0   1 mile

(Courtesy of the Grand Canyon Association, Geologic Map of the Grand Canyon by George Billingsly et al.)

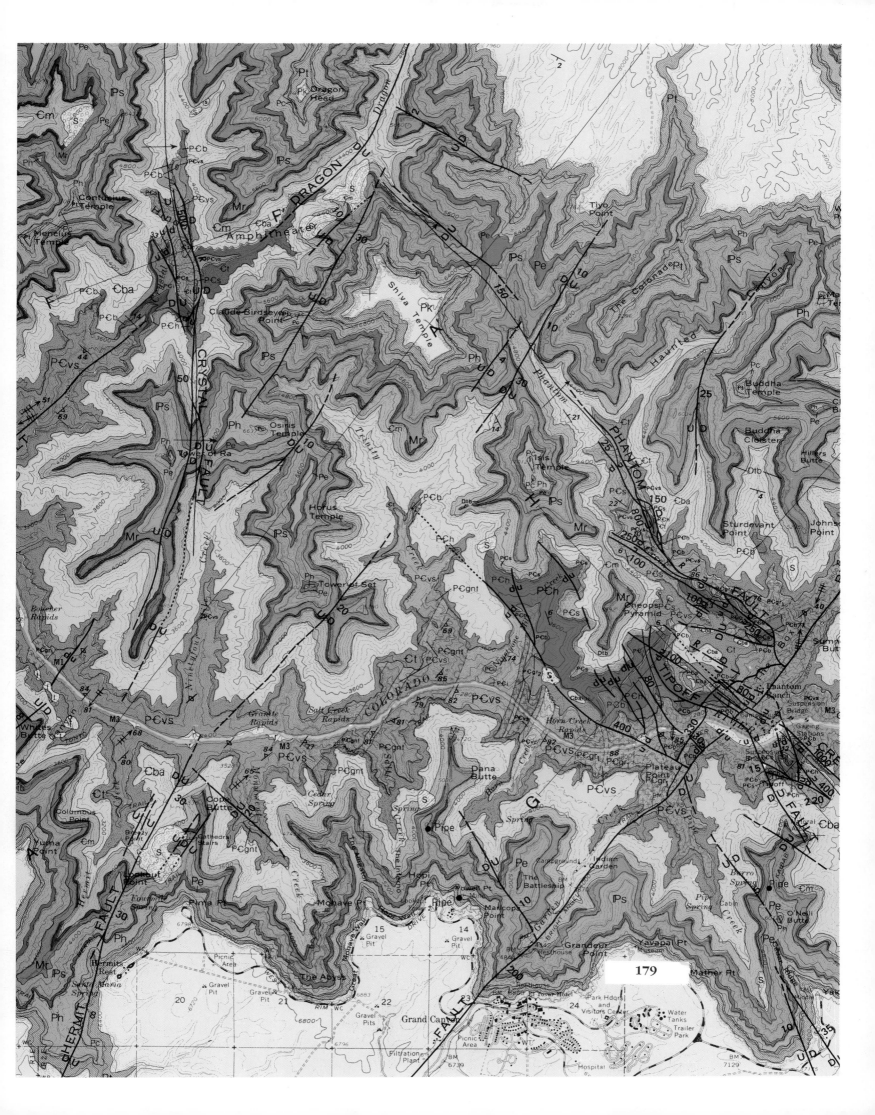

179

18. Geologic Map of Michigan and the Surrounding Area

The maps on pages 180–185 are parts of the geologic map of the United States published by the U.S. Geological Survey at a scale of 1 inch equals 25 miles. Rather than being mapped according to specific rock formations, as is the custom for larger-scale geologic maps, the rock units are mapped according to age.

a. Where are the youngest rocks exposed in this area?

b. Study the patterns made by the rock units in Michigan and in the adjacent areas. What major structural feature is shown?

c. Note the age of the deformed units in this structure. What is the oldest age assignable to the deformation?

d. Assuming that the units on this map dip at a low angle, sketch an east–west geologic cross section through the center of Michigan.

e. Around the Illinois–Indiana state line, note the outcrop pattern of the lower Paleozoic rocks (Ordovician and Devonian) and that of the upper Paleozoic rocks (Carboniferous) in the area. What major structural feature is present in this area?

f. What type of fault (normal, low-angle thrust, or strike-slip) is evident south of Saginaw Bay? What was the relative movement—north side up or down?

g. What major unconformities occur in the rock sequence in this area?

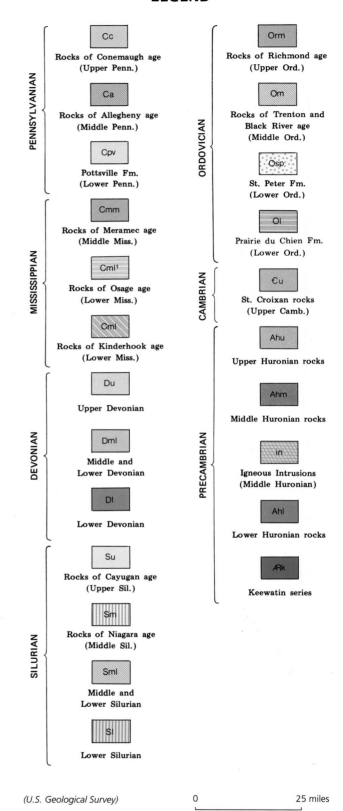

LEGEND

PENNSYLVANIAN
- Cc — Rocks of Conemaugh age (Upper Penn.)
- Ca — Rocks of Allegheny age (Middle Penn.)
- Cpv — Pottsville Fm. (Lower Penn.)

MISSISSIPPIAN
- Cmm — Rocks of Meramec age (Middle Miss.)
- Cml¹ — Rocks of Osage age (Lower Miss.)
- Cml — Rocks of Kinderhook age (Lower Miss.)

DEVONIAN
- Du — Upper Devonian
- Dml — Middle and Lower Devonian
- Dl — Lower Devonian

SILURIAN
- Su — Rocks of Cayugan age (Upper Sil.)
- Sm — Rocks of Niagara age (Middle Sil.)
- Sml — Middle and Lower Silurian
- Sl — Lower Silurian

ORDOVICIAN
- Orm — Rocks of Richmond age (Upper Ord.)
- Om — Rocks of Trenton and Black River age (Middle Ord.)
- Osp — St. Peter Fm. (Lower Ord.)
- Ol — Prairie du Chien Fm. (Lower Ord.)

CAMBRIAN
- Cu — St. Croixan rocks (Upper Camb.)

PRECAMBRIAN
- Ahu — Upper Huronian rocks
- Ahm — Middle Huronian rocks
- in — Igneous Intrusions (Middle Huronian)
- Ahl — Lower Huronian rocks
- Ark — Keewatin series

(U.S. Geological Survey)

0   25 miles

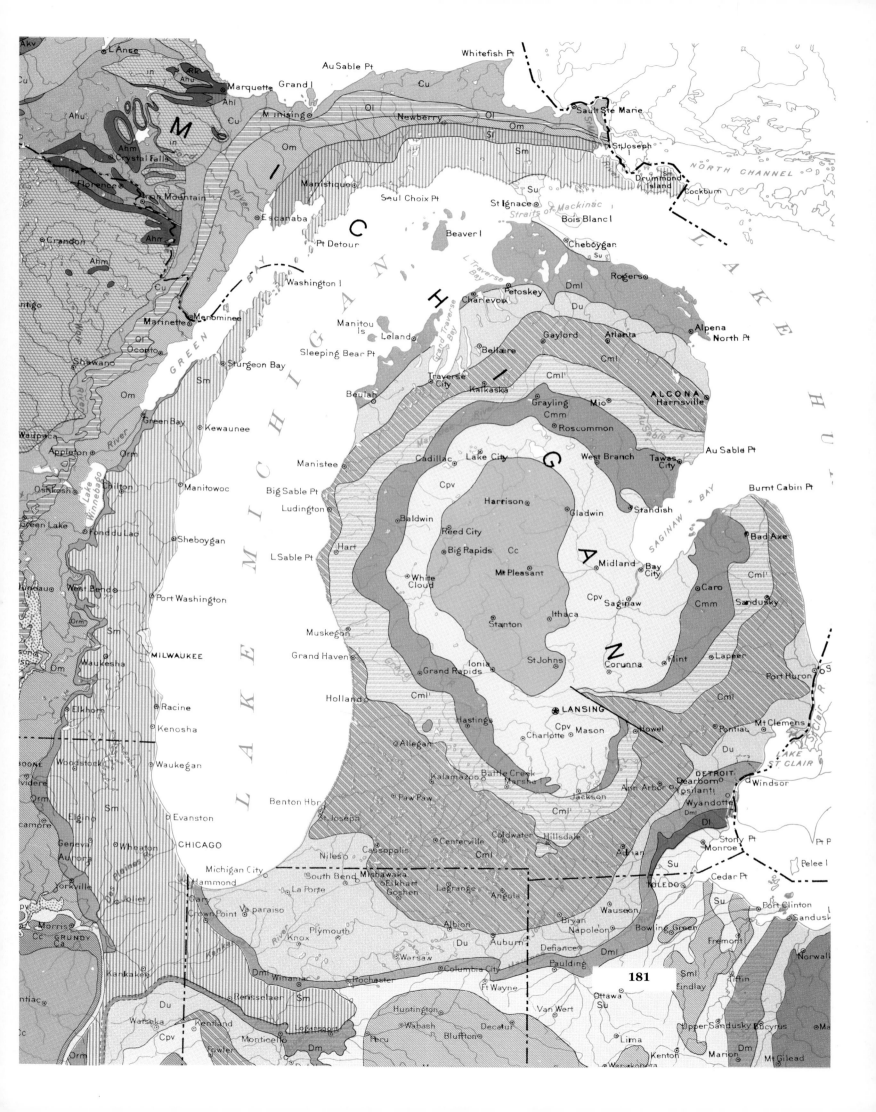

19. Geologic Map of the Wyoming Area

a. Study the outcrop pattern in the northeastern quarter of the map. This area encompasses the Black Hills of South Dakota and Wyoming. Describe the geologic structure.

b. What is the age of the oldest rocks in the Black Hills?

c. Is the structure of the Black Hills symmetric or asymmetric? On what facts did you base your answer?

d. Draw a geologic cross section along an east–west line through the center of the Black Hills and show all beds at the surface and in the subsurface.

e. Judging from the outcrop pattern of the White River Group (φw) on the east flank of the structure, what would be the age of the deformation in this area?

f. Are the igneous intrusions older or younger than the formation of the sedimentary rocks in this area? What map patterns support your answer?

g. What major structure is located between the Black Hills and the town of Buffalo, Wyoming?

h. Study the outcrops of the Precambrian rocks in Wyoming and Colorado. What structures are associated with these outcrops?

i. Study the outcrops of Tertiary rocks in Wyoming, especially the Wasatch (Ews), Fort Union (Efu), and Lance (El) formations. What structures are indicated by the outcrops of these formations?

j. Using a red pencil or felt-tipped pen, complete the geologic map by drawing the hinge line of the major domes and basins.

k. Study the outcrop patterns of the White River Group (φw) throughout Wyoming. What do these patterns tell you about the age of the deformation of the Rocky Mountains?

LEGEND

QUATERNARY

Qa — Alluvium and windblown sand

QPv — Younger volcanic rocks

TERTIARY

Miocene

Mbb — Browns Peak and Bishop Fms.

Moa — Ogallala and Arikaree Fms.

Oligocene

φw — White River Gr.

Eocene

Eb — Bridger Fm.

Egr — Green River Fm.

Ti — Intrusive rocks

Ews — Wasatch Fm.

Efu — Fort Union Fm.

El — Lance Fm.

CRETACEOUS

Km — Montana Gr. (Upper Cretaceous)

Kmv — Mesaverde Gr. (Upper Cretaceous)

Kc — Colorado Gr. (Upper Cretaceous)

Kdl — Dakota Fm. and Lower Cretaceous rocks

JURASSIC

KJ — Dakota to Morrison Fms. (Cretaceous to Jurassic)

J — Jurassic rocks

TRIASSIC

JTR — Jurassic and Triassic rocks

TR — Triassic rocks

PERMIAN

Cm — Lower Permian rocks

PENNSYLVANIAN

Cpp — Permian and Pennsylvanian rocks

Cp — Pennsylvanian rocks

LOWER PALEOZOIC

DC — Devonian to Cambrian rocks

CO — Ordovician to Cambrian rocks

PRECAMBRIAN

Ai — Intrusive rocks

As — Metamorphic rocks

PRg — Granite and other intrusive rocks

PR — Metamorphic rocks

(U.S. Geological Survey)

0   25 miles

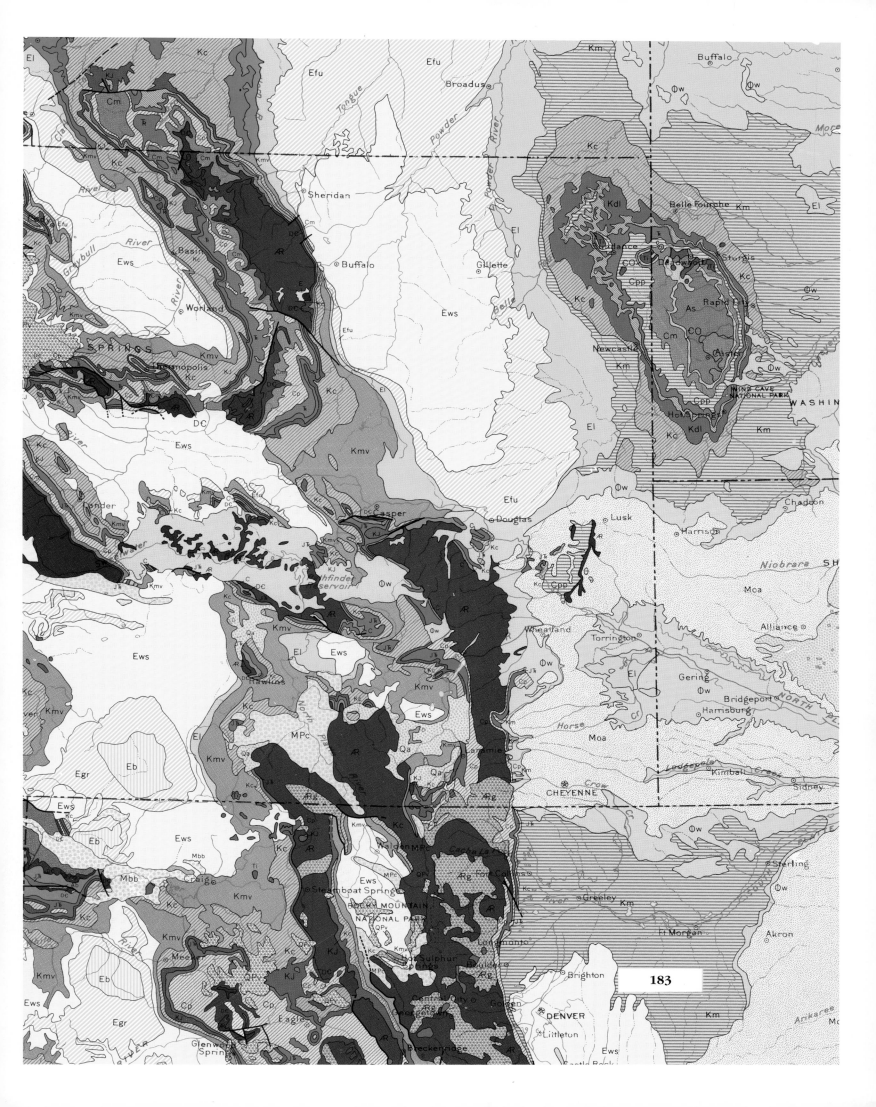

183

20. Geologic Map of the Southern States

a. What is the regional strike of the Cretaceous and Tertiary rocks in southern Alabama?

b. In what direction do these rocks dip?

c. What is the regional strike of the Paleozoic rocks in Tennessee, Georgia, and northeastern Alabama?

d. What structural features are indicated by the outcrop patterns of these Paleozoic rocks?

e. What is the structural relationship of the Cretaceous rocks and the underlying Paleozoic rocks?

f. If you were to divide the area into geologic regions, where would you draw the boundaries between major provinces? On what factors did you base your decision?

g. Outline a sequence of events to account for the differences in outcrop patterns and age relationships in the area.

h. Are the faults in eastern Tennessee normal or thrust faults?

i. From the structural trend of the Paleozoic rocks, determine the orientation of the forces responsible for pre-Cretaceous deformation.

j. What structural feature dominates the northwest quarter of the map?

LEGEND

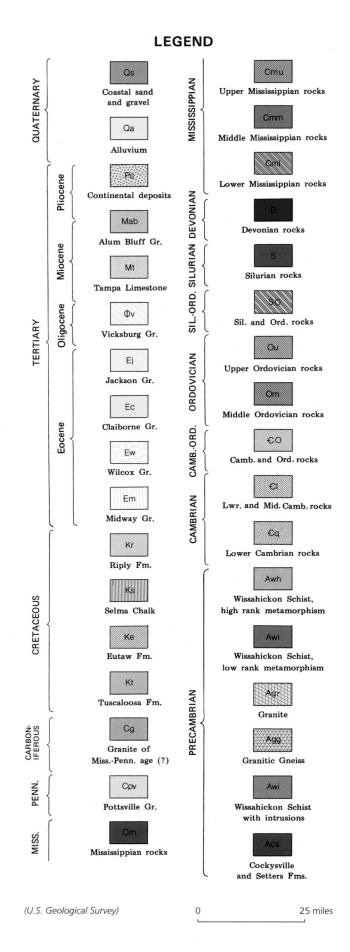

(U.S. Geological Survey)

0   25 miles

185

21. Geologic Map of Chico, California

Although the patterns of the rock bodies shown on this map appear complex, age relationships can be determined by studying the nature of the rock boundaries (contacts). Keep in mind the law of crosscutting relationships, and the general shape of intrusive rocks (plutonic rocks) and metamorphic rocks. It may be a good idea to review again the general outcrop pattern of intrusive rock bodies (Figure 16.11) and the outcrop patterns of surficial deposits (Figure 16.12).

a. What map evidence indicates that the plutonic rocks are younger than the metamorphic rocks?

b. How many different periods of intrusion are shown on this map?

c. What types of faults occurred in this area?

d. List the major types of surficial deposits.

e. What map evidence indicates that the extrusive rocks are younger than the metamorphic and intrusive rocks?

f. What is the relationship between the Lovejoy Basalt and the present drainage system?

g. What map evidence indicates the relative age of the Lovejoy Basalt and the pyroclastic andesites?

h. How do the outcrop patterns of the major igneous intrusions compare with those shown in Figure 16.11?

SURFICIAL DEPOSITS

Qgr — Gravel deposits

EXTRUSIVE ROCKS

Tbo — Olivine basalt

Ta — Pyroxene andesite

Tpa — Pyroclastic andesite

Tl — Lovejoy basalt

PLUTONIC ROCKS

Kltr — Trondhjemite

KJqd — Quartz diorite/tonalite

KJhd — Hornblende-biotite quartz diorite

KJpd — Pyroxene diorite

Jgb — Gabbro

METAMORPHIC INTRUSIVES

sp — Serpentinite-talc schist

mtr — Metatrondhjemite

mdi — Metadiorite

mgb — Metagabbro

mqp — Metaquartz porphyry

hgn — Hornblende gneiss

METAMORPHIC ROCKS

mb — Metabasalt

ma — Meta-andesite

md — Metadacite

mr — Metarhyolite

mt — Metatuff

mm — Marble

mq — Quartzite-metachert

mp — Phyllite

FORMATIONS

 FRANKLIN CANYON FORMATION

BLOOMER HILL FORMATION

DUFFEY DOME FORMATION

—— FAULT

CI 50 m

0   2 miles

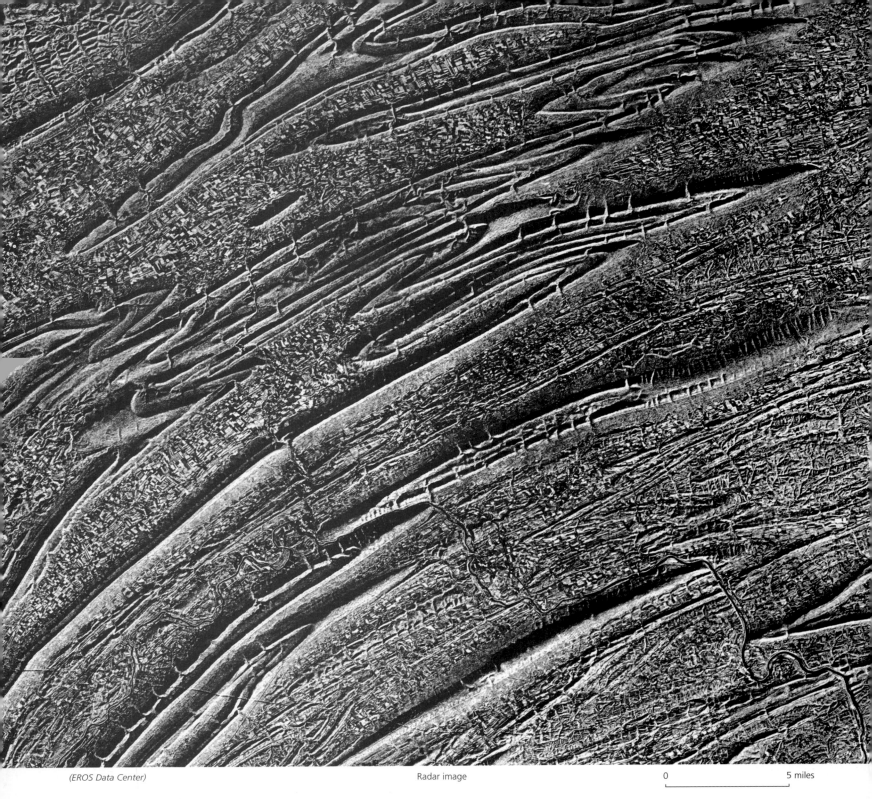

Radar image 0 5 miles

22. Radar Image of Central Pennsylvania

This is a radar image of part of the Valley and Ridge province of the Appalachian Mountains. Because resistant sandstones stand up as distinctive ridges, it is possible to make a preliminary geologic map of the area by tracing the ridges.

   a. Draw a preliminary geologic map of the area by tracing the extent of the ridge-forming formations.

b. Identify the major anticlines and synclines, and complete the geologic map by adding the proper symbols for fold hinge lines. Show the direction of plunge.

c. Identify the minor folds (both anticlines and synclines). How are the minor folds related to the major folds, with respect to their orientation (trend of axial trace) and direction of plunge?

d. Study the large anticlines shown on the southern part of the map.

How can you identify these folds as anticlines?

e. What was the orientation of the force that deformed this sequence of rocks?

f. Where are the oldest rocks exposed in this area?

g. Draw an idealized north–south cross section through the map.

h. Develop a legend similar to those on pages 178–187 to explain this map.

(Earth Satellite Corporation)

0   10 miles

23. Landsat Image of San Rafael Swell—Central Utah

The desert regions of the Colorado Plateau have little soil and vegetation, so the color and tone of the Landsat image of the area are produced largely by the coloration of the exposed rock. The image has been enhanced by computer. The false colors are exaggerated to emphasize the contrasts, almost making the image itself a geologic map. The brilliant pink indicates areas covered with vegetation.

a. Draw a geologic map of the area by tracing the contacts of the major rock units indicated by different colors and by different topographic expressions.

b. What geologic structure dominates this area?

c. How does this structure compare with that of the Black Hills, shown on the map on page 183?

d. Where are the oldest rocks exposed?

e. Where are the youngest rocks exposed?

f. What features produce the linear texture (colored yellow) in the lower-right corner of the image?

g. If you were exploring for petroleum and natural gas, where in this region would you recommend drilling?

h. What types of surficial deposits occur in this area?

189

EXERCISE 17 SEISMOLOGY

OBJECTIVE

To understand how the study of seismic waves is used to determine Earth's internal structure.

MAIN CONCEPT

Changes in direction and variations in velocity with which seismic waves pass through Earth are evidences of distinct discontinuities in internal structure and give us a basis for interpreting Earth's internal structure.

SUPPORTING IDEAS

1. Seismic reflection profiles provide an effective way to study relatively shallow subsurface structure and stratigraphy.
2. The shadow zone of both the P and S waves indicates the size of Earth's core and provides important information about the state of matter (solid or liquid) in the core and mantle.

The exploration of the subsurface structure of Earth by monitoring seismic waves has been possible since the 1920s. With the recent advent of solid-state electronic components, the development of magnetic tape records, and the use of computers, seismic exploration has reached a high level of sophistication. The seismic records shown in Figure 17.1 are examples of the types of seismic data that geologists commonly analyze. Each vertical line is a record of energy received at a recording station.

When the energy is from a reflective horizon, such as a resistant sandstone formation, it is recorded as a positive or negative segment of the curve (deviation to the right or left, respectively). Positive reflections, which occur when seismic energy encounters a higher velocity layer below a lower velocity layer, are generally shaded black for emphasis. This is quite clear in Figure 17.1A, which shows small-scale detail from a single recording.

When a series of seismic records are made along a line, they show the configuration of the rock units in the subsurface and can be interpreted in much the same way as geologic cross sections (Figures 17.1B–17.1D). (There are, however, a number of geophysical limitations that must be considered by the geologist, but for our purposes, we will interpret the

seismic sections as we would interpret a geologic cross section.)

Figure 17.2 illustrates the major steps involved in the interpretation of seismic records. The uninterpreted record (Figure 17.2A) shows the major geologic elements, but their boundaries are fuzzy, like a poor image on a TV screen, with a lot of interference ("snow"). Preliminary interpretation involves highlighting in color the more prominent units, which are semicontinuous across the section. The units we have colored yellow, red, green, and blue are the most obvious (Figure 17.2B). Discontinuities exist in all of the units. In all probability, several of them are faults. The most obvious discontinuities are marked with a heavy black line. Our highest confidence level is at this stage of interpretation.

We now elaborate on our first interpretation (Figure 17.2C). Judging from what we have mapped, we can extend or project the units we have identified into areas that are somewhat less certain. Based on our understanding of geologic strata, faults, and folds, we can extend our interpretation across the entire section. Some areas undoubtedly will remain speculative, and it may be wise to consider several alternative possibilities. To do this, make several copies of the uninterpreted record and work up different interpretations of the data.

PROBLEMS

This exercise involves a variety of examples of how geologists use seismicity in studying the internal structures of Earth. We will first interpret seismic records of shallow structures and then determine the depth of Earth's core. As a final exercise, we will work on a hypothetical problem of determining the size of the core in planet Mars.

A. A Record from a Single Recording Station shows variations in wave velocity. The positive segments of the curve (deviations to the right) result from strata that transmit seismic waves at a high velocity. These segments are rendered in black for emphasis.

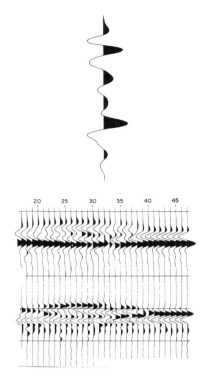

B. A Number of Recording Stations Positioned in a Line produced this seismic record. The positive segments of the record outline the boundaries between rock units that have different seismic characteristics (for example, a unit of sandstone and a unit of shale).

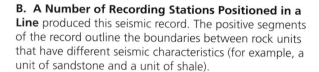

C. A Line of Seismic Stations may record a profile of the major rock bodies.

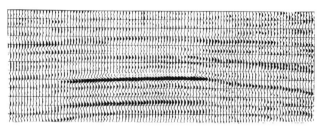

D. A Longer Line of Seismic Stations may record the subsurface structure of the rock units.

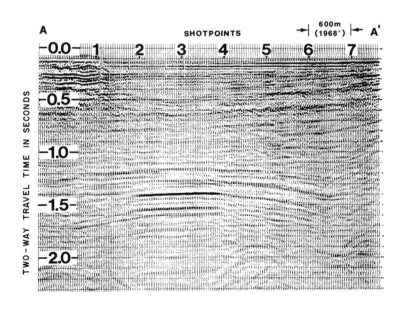

FIGURE 17.1
Examples of Seismic Records
(American Association of Petroleum Geologists)

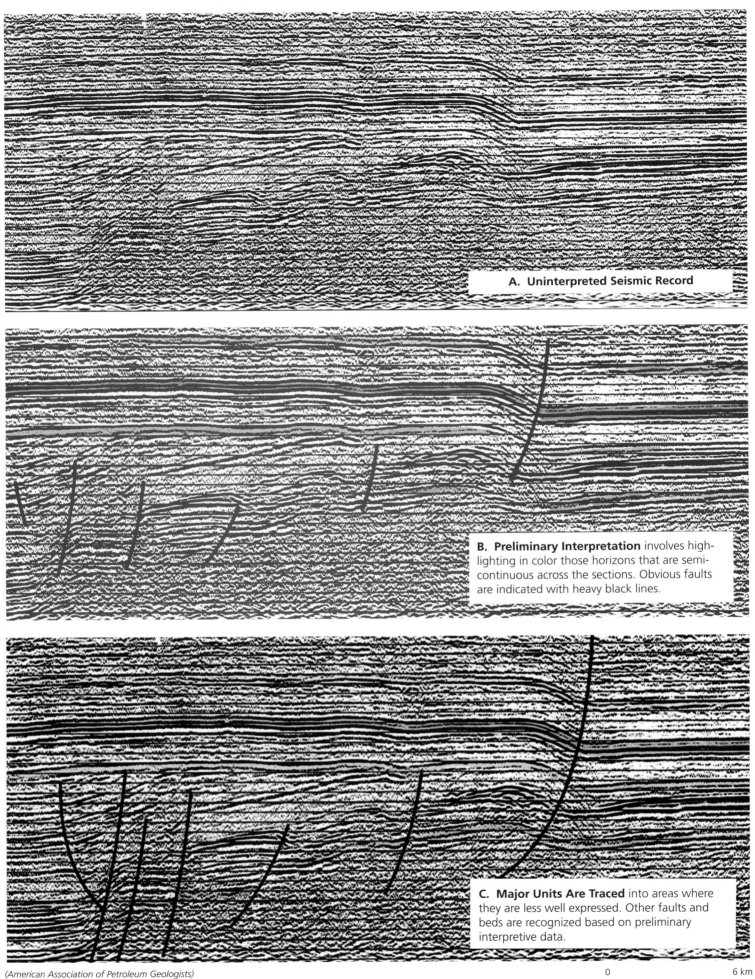

A. **Uninterpreted Seismic Record**

B. **Preliminary Interpretation** involves highlighting in color those horizons that are semi-continuous across the sections. Obvious faults are indicated with heavy black lines.

C. **Major Units Are Traced** into areas where they are less well expressed. Other faults and beds are recognized based on preliminary interpretive data.

(American Association of Petroleum Geologists)

0   6 km

FIGURE 17.2
Interpretation of Seismic Records

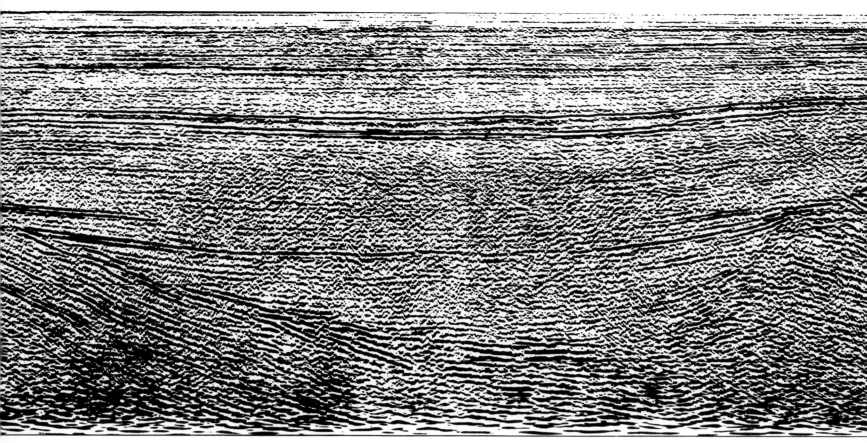

(American Association of Petroleum Geologists)

1. Seismic Section for Interpretation Problem

a. Interpret this seismic section by sketching the major boundaries between reflective horizons and by coloring the rock units between the boundaries in different colors. Using a colored pencil, lightly sketch those horizons that are semi-continuous across the section and that separate rock units of similar inclination. (See Figure 17.2B.) Then sketch in the inferred faults, which are expressed by abrupt discontinuities in the major units (Figure 17.2B). Complete the section by extending the units into areas where they are less well expressed (Figure 17.2C).

b. Outline the major geologic events as interpreted from the data on the seismic section.

2. San Francisco Bay Area

a. What controls the distribution of earthquakes in the San Francisco Bay area?

b. What was the magnitude of the largest earthquake recorded between 1972 and 1989?

c. What was the magnitude of the average earthquake recorded between 1972 and 1989?

d. The San Andreas Fault is a strike-slip fault that extends across the area west of San Francisco. What evidence of the fault extends out into the ocean, just west of the inlet to San Francisco Bay?

e. What is the evidence of the fault north of San Francisco Bay?

f. Trace the major faults in the area as they are expressed at the surface. Are all seismically active?

g. Much of the shoreline around the bay area is a tidal flat, or was a tidal flat in recent geologic time. How does this fact influence the potential destruction that could occur from an earthquake in that area?

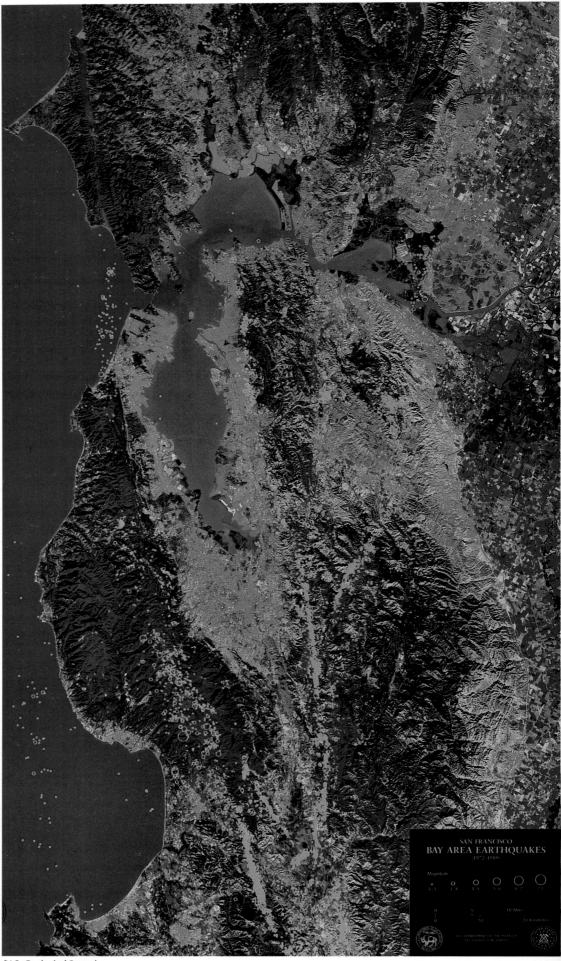

(U.S. Geological Survey)

3. The Internal Structure of Earth

Two important facts permit us to determine the radius of Earth's core: (1) The S-wave shadow zone begins at 11,500 km (surface distance) from any earthquake epicenter, and (2) the radius of Earth is 6370 km.

a. Find the size (radius) of the liquid core by constructing a scale drawing.

b. Compare your result with the accepted value, indicated in your textbook. Explain possible reasons for any difference between your result and the accepted value.

c. Using the same general procedure, try the following problem: If Mars has a radius of 3400 km and has an S-wave shadow zone beginning at 7000 km, what is the size of its core?

Suggested procedure:

1. Use a drawing compass to draw a circle (to scale) representing Earth in cross section.

2. Pick any point on the perimeter of the circle and label it "Epicenter."

3. Plot the S-wave shadow zone on the circle as follows: The point on the perimeter of the circle where the shadow zone begins can be determined by measuring, on the circumference of the circle, the surface distance of 11,500 km.

4. Draw a line representing a seismic ray (a line drawn perpendicular to the wave front) between the epicenter and the beginning of the shadow zone. This line, in effect, defines the outer boundary of the shadow zone that S waves cannot enter.

5. Repeat this procedure for several different epicenters.

6. Now, use a drawing compass to draw in Earth's core—a circle that is tangent with the seismic rays that define the shadow zone.

7. Measure the radius of the circle representing the core. (By using simple trigonometry, the size of the core can be obtained by calculation rather than by measurement.)

4. Radar Interferometry and Earthquake Deformation

If two radar images are made from a satellite from exactly the same position, and if the Earth surface changes ever so slightly between the radar scans, a map of the differences called an interferogram is displayed as a series of colored "fringes" that are like contours on a topographic map. Each fringe marks 10 cm of vertical ground motion. By starting at the edge of the image and counting the number of colored bands and multiplying by 10 cm, you can determine the amount of deformation throughout the area.

a. Construct a contour map showing deformation of the area by counting the fringes.

b. What was the maximum deformation (vertical uplift)?

c. What controlled the areas of maximum uplift?

d. Why is the deformation concentrated in some areas and not others?

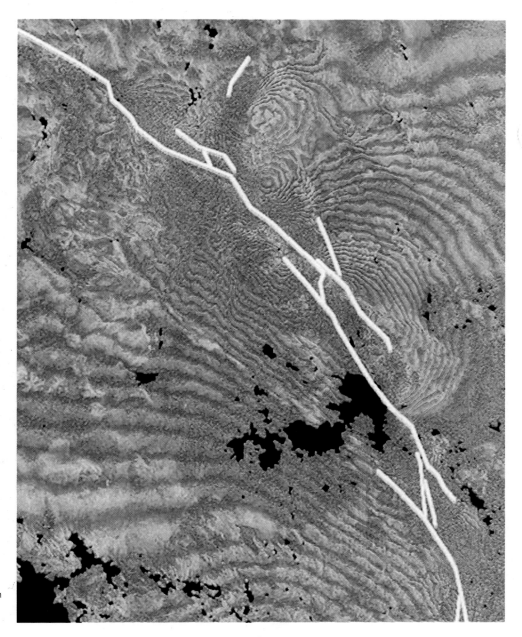

(NASA - JPL)

0   10 km

One color cycle = 2.8 cm of range change

EXERCISE 18 — PLATE TECTONICS

■ OBJECTIVE

To become aware of the location and topographic expression of the major tectonic features of Earth and to understand how these features are related to the plate tectonic theory.

■ MAIN CONCEPT

The theory of plate tectonics explains the origin and evolution of Earth's structural features as the result of a system of moving lithospheric plates. This motion involves sea-floor spreading, continental drift, and subduction. Plate tectonics results in the creation of new oceanic crust, the growth of continents, volcanism, earthquakes, and mountain building.

■ SUPPORTING IDEAS

1. Plate boundaries are the most dynamic zones of Earth's lithosphere.
2. The major components of all continents are (a) shields, (b) stable platforms, and (c) folded mountain belts.
3. Plate motion is indicated by the orientation of most major structural and topographic features, such as the mid-oceanic rift, fracture zones, and mountain belts.
4. Many local characteristics (e.g., island arcs, mountain belts, volcanic activity, and features on the ocean floor) can be explained by specific conditions of their tectonic setting.

THEORY OF PLATE TECTONICS

The theory of plate tectonics provides a master plan of Earth's crustal dynamics—a framework in which numerous observations of Earth are united into a comprehensive whole. The great power of this theory is that it can explain, with equal ease, such diverse geologic phenomena as mountain building, the evolution of ocean basins, the chemistry of lavas, rates of sedimentation, and the migration and extinction of plants and animals. Plate tectonics explains Earth's dynamics—the way it works—and is probably the most significant scientific breakthrough in the history of geology.

The basic elements of the plate tectonics theory are simple and can be easily understood by carefully studying the diagram in Figure 18.1. The lithosphere, which includes Earth's crust and part of the upper mantle, is rigid, but the underlying asthenosphere flows slowly. The fundamental idea of plate tectonics is that the segments, or plates, of the rigid lithosphere are in constant motion relative to one another and carry the lighter continents with them.

The lithospheric plates form as hot mantle material rises along midoceanic ridges and are destroyed in subduction zones, where one of the converging plates plunges into the hotter mantle below (Figure 18.1). Their descent is marked by deep-sea trenches. Where plates slide by one another, large fracture systems occur. The movement and collision of plates account for most of Earth's earthquakes, volcanoes, and folded mountain belts, as well as for the drift of its continents.

Plate boundaries do not necessarily coincide with continental boundaries, although some do. There are seven very large plates and a dozen or more smaller plates. Each plate is about 100 km thick. Plates generally move at rates between 1 and 10 cm/yr, although rates of interactions between adjacent plates may exceed 20 cm/yr. Because the plates are internally rigid, they mostly interact along their edges.

The theory of plate tectonics grew out of our newly acquired abilities to study the geology of the ocean floors, measure rock magnetism, and study Earth's seismicity on a global scale. The vast quantities of data generated by these studies have been recorded on a variety of maps. In this way, the data can be synthesized, analyzed, and interpreted. One of the most useful maps for this purpose is the computer-generated physiographic map of the world (pages 198–199).

PROBLEMS

The map on pages 198–199 shows the ocean floor and the continents in color-coded shaded relief. The following questions are involved largely with the study of this map—plotting geologic data and interpreting the tectonic relationships of Earth's features on a global scale. The map on pages 198–199 was generated by computer, using elevations measured by satellites.

1. Draw the plate boundaries on the map. Use a soft pencil at first, and when you are satisfied with your work, emphasize the boundaries with a light-colored felt-tipped pen. Show convergent, divergent, and passive boundaries with different colors.

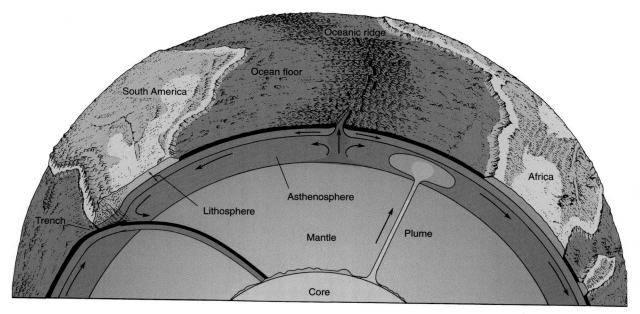

FIGURE 18.1

The Tectonic System operates from Earth's internal heat. The asthenosphere is more plastic than both the overlying lithosphere and the underlying lower mantle because of an optimum balance of temperature and pressure. Above the plastic asthenosphere, relatively cool and rigid lithospheric plates split and move apart as single mechanical units. Molten rock from the asthenosphere wells up to fill the void and creates new lithosphere. Convection circulation in the athenosphere moves away from the spreading center to areas where plates descend into the lower mantle at oceanic trenches. Some plates contain blocks of thick, low-density continental crust that cannot sink into the mantle. As a result, where a plate carrying continental crust collides with another plate, the continental margins are deformed into mountain ranges. The plate margins are the most active areas on Earth—the sites of the most intense volcanism, seismic activity, and crustal deformation.

2. List the features shown on the map that indicate plate motion.

3. Show the direction of plate motion with a series of arrows.

4. Draw an idealized cross section through the plate margins that occur near (a) South America, (b) Central America, (c) Japan, (d) the Himalaya Mountains, and (e) the Indonesian islands.

5. On the basis of the information you have plotted, explain briefly the origin of (a) the Red Sea, (b) the Black Sea, (c) the Mediterranean Sea, and (d) the Gulf of California.

6. Why are the Himalaya Mountains higher than the Andes Mountains?

7. What are the relationships among the islands of Indonesia, the Himalayas, and the Alps? Consider such features as the nature of the converging plates and the characteristics of a subduction zone.

8. If you were involved with a deep-sea drilling project, where would you suggest drilling to find the oldest sediments in the Pacific Ocean?

9. The average rate of plate motion is about 3.8 cm/yr. Assuming that this rate of plate movement has been constant, how old is the Atlantic Ocean?

10. Explain the origin of the Cascade Mountains in Oregon and Washington.

11. Note the similarities and differences between these two tectonic settings: (a) the area between Florida and South America (Caribbean Sea) and (b) the area between South America and Antarctica (Scotia Sea). From your observations, do you conclude that these areas had similar or different origins?

12. Is the southern part of Central America most likely to be (a) a segment of a continental shield, (b) part of a stable platform, (c) a former volcanic arc, or (d) part of a mountain belt of folded sedimentary rocks?

13. Australia and New Guinea are part of the same continental block and are separated from each other by a very shallow sea. What is the origin of the mountains of New Guinea?

14. North and South America, Australia, and India all have mountain belts along one of their margins. Why are there no major mountain belts along the east, south, and west coasts of Africa?

15. Why are Asia, Europe, and India considered in the geologic sense to be three independent continental masses?

16. Africa is a relatively high continent; large areas of its shield are more than 6000 ft above sea level. On most other continents, the shield is near sea level. How might this phenomenon be explained in light of Africa's unique relationship to the oceanic ridge system?

FIGURE 18.2
Digital Color Shaded Relief Map of the World. The surface expression of plate boundaries is clearly expressed by variations in elevation of Earth's surface. Divergent plate boundaries are expressed by the mid-oceanic ridge. Transform plate boundaries and associated fracture zones are expressed by offset in the oceanic ridge. Convergent plate boundaries are expressed by active folded mountain belts and by deep sea trenches and associated island arcs.

(Ken Perry, Chalk Butte, Inc.)

NOTE: Please refer to the map on the inside back cover for geographic locations and names of features referred to in this and subsequent exercises.

17. Tectonic Map of North America

a. Study the structural patterns of the shield. What evidence on the map indicates that the shield consists of the roots of a series of deformed mountain belts? Indicate on the map, with arrows, the direction of the compressive forces necessary to produce the structural trends in each major area of the shield.

b. If a mountain belt results from the collision of two lithospheric plates, how many major mountain-building pulses are apparent in the Canadian shield?

c. How does the radiometric age of the rocks in the shield support the concept of continental growth by accretion?

d. Note the thickness of the sedimentary rock cover of the stable platform, indicated by contour lines. How thick is this rock cover throughout most of the platform?

e. Explain why the sedimentary rock cover of the stable platform was probably deposited in very shallow water rather than in the deep ocean.

f. How are the orientation and location of the mountain systems of North America explained by the theory of plate tectonics?

g. The reddish-orange colors on the map represent granitic intrusions into young, folded mountain belts. Use the theory of plate tectonics to explain how the intrusions formed and how they are related to mountain building.

h. Explain the tectonic origin of Baja California. If the present tectonic forces continue to be active in the western United States, what will happen to Baja California and southern California?

i. Explain the origin of the volcanoes in the Cascade Mountains.

j. Note the submarine contours west of Central America. What tectonic features do they represent?

k. How does the theory of plate tectonics explain the concentration of volcanic activity along the west coast of Central America?

SEDIMENTARY UNITS

Thick deposits in structurally negative areas

Synorogenic and post-orogenic deposits

Miogesynclinal deposits

Eugeosynclinal desposits

Early geosynclinal deposits
Of Middle and Upper Proterozoic ages

Basement massifs
Mainly of Precambrian age. Includes metamorphic complexes that involve younger rocks

VOLCANIC AND PLUTONIC UNIT

Granitic plutons
Ages are generally within the span of the tectonic cycle of the foldbelts in which they lie

SPECIAL UNIT

Eugeosynclinal deposits of the Pacific border
Includes Franciscan Formation of California

PRECAMBRIAN FOLDBELTS

Dark colors show areas of paraschist and paragneiss derived from supracrustal rocks; light colors show areas of granite and orthogneiss of plutonic origin

Greenville foldbelt
Deformed 880-1,000 m.y. ago

Rocks of the Hudsonian foldbelt
Overprinted by Elsonian event about 1,370 m.y. ago

Hudsonian foldbelts
Deformed 1,640-1,820 m.y. ago

Kenoran foldbelts
Deformed 2,390-2,600 m.y. ago

Anorthosite bodies
In Greenville and Elsonian belts or, alternatively, in eastern Canadian Shield

PLATFORM AREAS

Ice cap of Quaternary age
On Precambrian and Paleozoic basement

Plateau basalts and associated rocks
In North Atlantic province

Platform deposits on Mesozoic basement
In Arctic Coastal Plain

Platform deposits on Paleozoic basement
In Atlantic and Gulf Coastal Plains

Platform deposits on Precambrian basement
In central-craton

Platform deposits within the Precambrian
Mainly in the Canadian Shield

STRUCTURAL SYMBOLS

⊢⊢⊢⊢⊢⊣ᵈ▵ᵈ▵
Normal fault
Hachures on downthrown side

⇌ ----
Transcurrent fault
Arrows show relative lateral movement

▲▲▲▲▲
Thrust fault
Barbs on upthrown side

—————————
Subsea fault
Long dashes based on topographic and geophysical evidence; short dashes, based on geophysical evidence only

Axes of sea-floor spreading

⊢——⊢——⊢—t—
Flexure
Arrows on depressed side

Salt domes and salt diapirs
In Gulf Coastal Plain and Gulf of Mexico

★
Volcano

———————— +2000
———————— 0
———————— 2000

Contours on basement surfaces beneath platform areas
All contours are below sea level except where marked with plus symbols. Interval 2,000 meters

(U.S. Geological Survey)

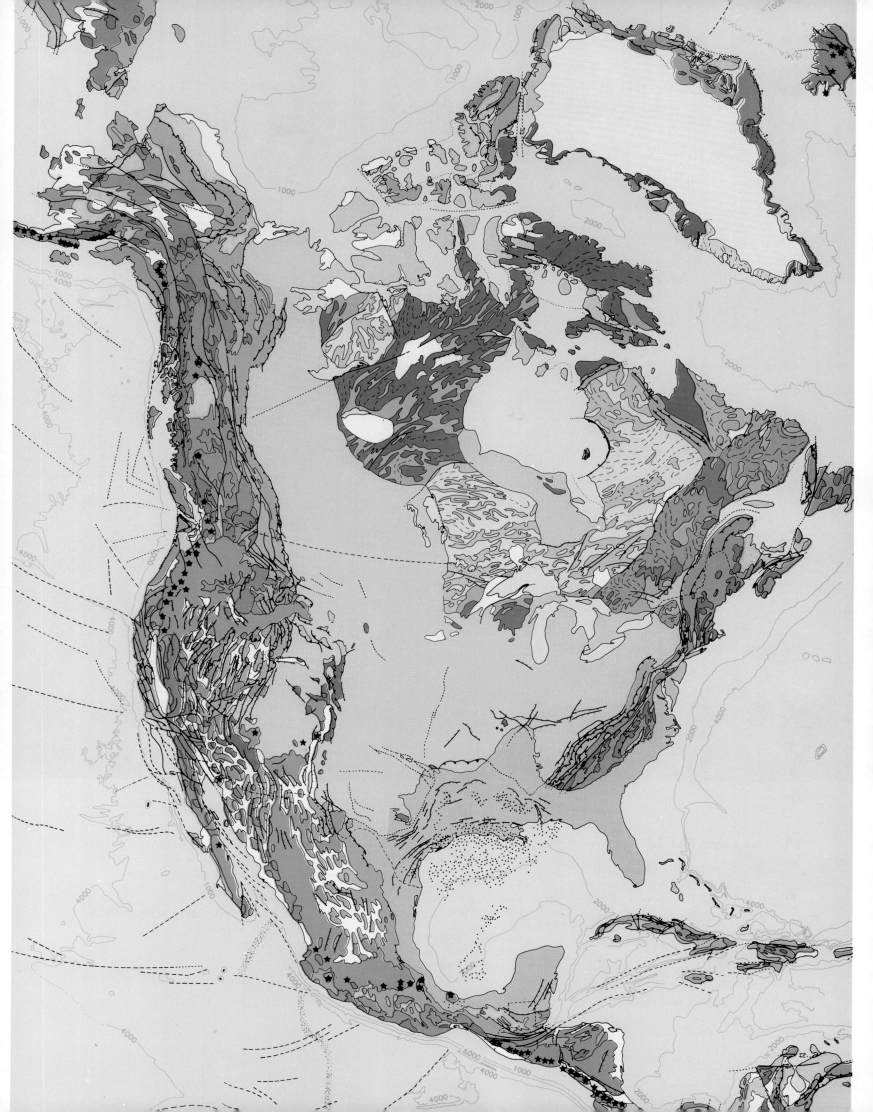

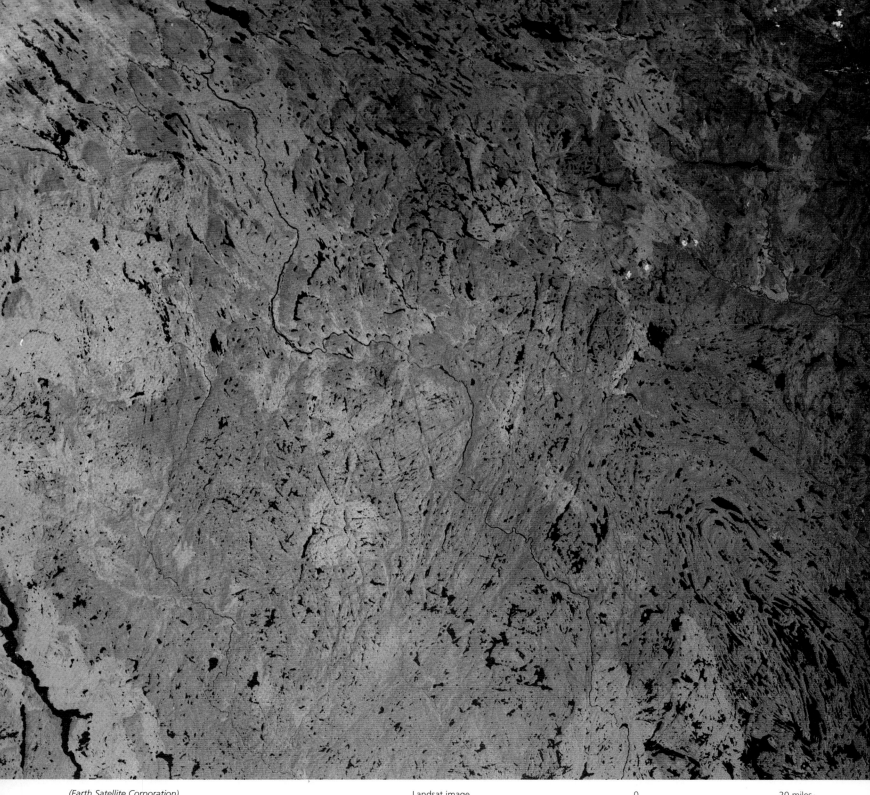

(Earth Satellite Corporation)

Landsat image

0   20 miles

18. Canadian Shield

a. Study the Landsat image and determine the general shape of the granitic bodies in this part of the shield. What type of igneous intrusions are they?

b. Study the structural trends (commonly expressed by linear lakes) of the metamorphic rocks. What evidence can you find that these rocks have been intensely deformed by horizontal stresses in the crust?

c. What geologic features indicate that the rocks in the shield are the roots of ancient mountains?

d. Why might you conclude that great volumes of rock have been eroded to form the shield?

e. Draw an idealized cross section showing the general characteristics of the structure and rock types in this area. Is this area typical of shields in general? (See Figure 16.11.)

f. Why are there so many lakes in this region?

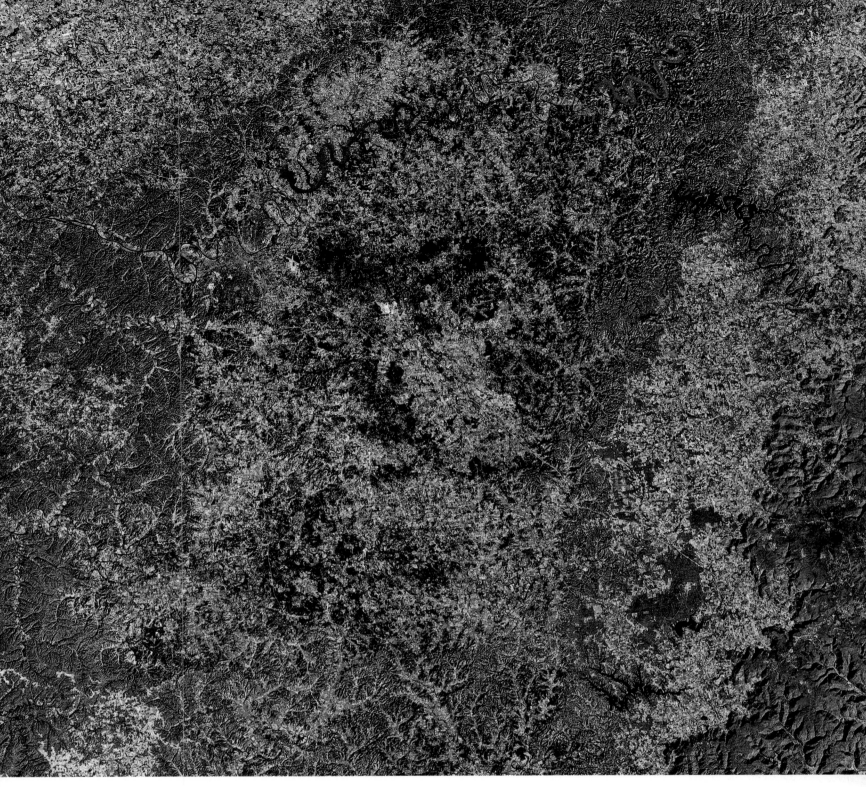

(Earth Satellite Corporation)

Landsat image

0   20 miles

19. Stable Platform—Kentucky

a. The Nashville Dome is typical of the broad structural upwarps of a stable platform. The crest of the dome has been eroded, so on a regional scale the outcropping rock formations display an elliptical pattern (see Figure 16.5). The rocks only dip a few degrees, however, so on a more local scale the contacts between formations have a dendritic pattern. One formation is expressed in a relatively dark-brown tone on this Landsat image and has a distinctive texture produced by a fine drainage network. Trace the boundaries of this formation on the image. The youngest formation exposed in the lower left part of the map is also expressed by a darker brown color but has a coarse drainage pattern. Draw the basal contact of the youngest formation. Next, draw a generalized cross section of the dome to show the structure's broad configuration and the surface expression of the rock formations.

b. How does the deformation of the stable platform differ from that of mountain belts?

c. Compare the landforms shown on the Landsat image of a shield (page 202) with the landforms shown on this image of the stable platform. What is the major difference between the two areas?

d. Why are rocks of the stable platform predominantly shallow-marine sediments?

203

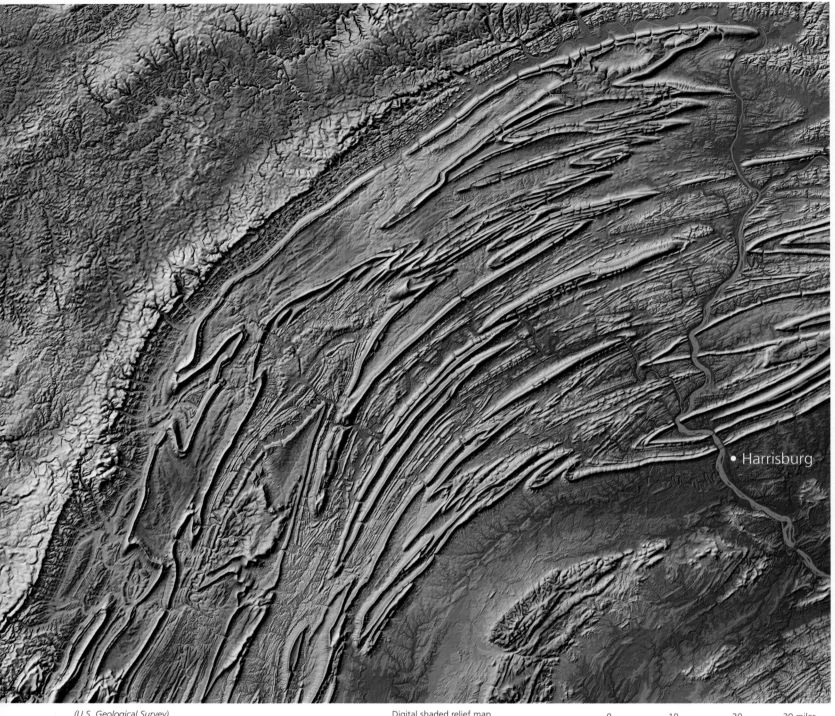

(U.S. Geological Survey)　　　　　　　　Digital shaded relief map　　　　　　0　　　10　　　20　　　30 miles

• Harrisburg

20. Folded Appalachian Mountains

In this area, the entire sequence of sedimentary rocks (nearly 40,000 ft thick) has been deformed into a series of tight folds. Erosion has sculpted the resistant formations into prominent ridges that zigzag across the landscape. The low, intervening valleys are eroded in nonresistant shale and limestone formations.

a. Study the image and note the configuration of the prominent ridges. Trace their extent across the area with a felt-tipped highlighter pen. Refer to the map patterns of plunging folds (see Figure 16.7), and label the anticlines and synclines.

b. Draw an idealized cross section of the folded belt to show the style

and degree of deformation. Show the folds below the surface with solid lines and show the parts of the folds that have been eroded away (above the surface) with dashed lines.

c. Indicate, with arrows, the direction of the stresses that produced this deformation.

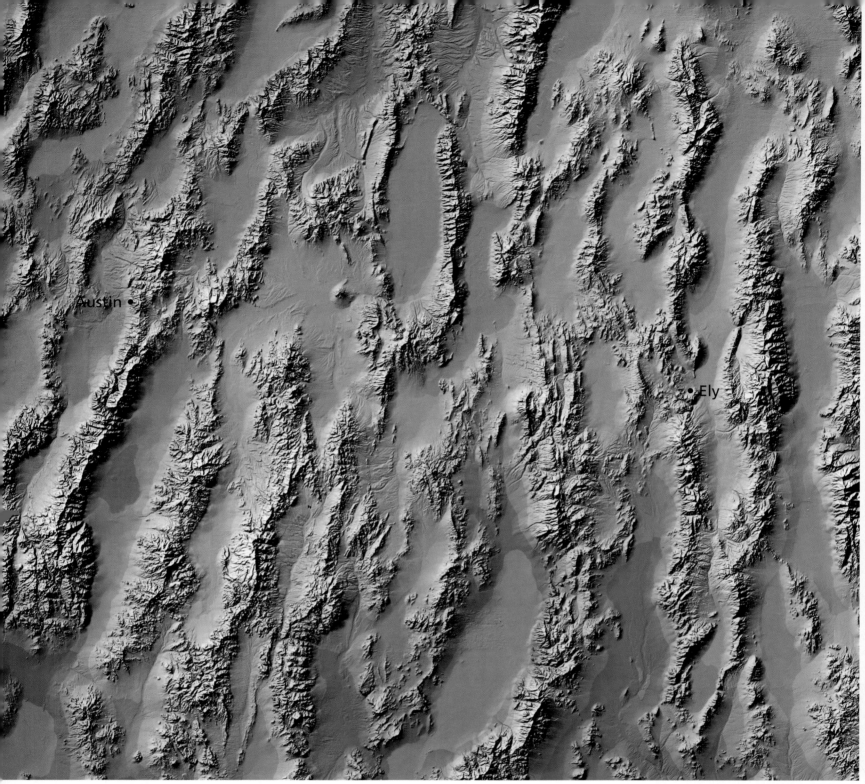

(U.S. Geological Survey)   Digital shaded relief map   0   50 miles

21. Basin and Range—Nevada

The Basin and Range province is a large region in the western United States. It consists of numerous mountain ranges that are mostly the result of extension and block faulting, separated by basins filled with erosional debris derived from the mountain ranges. It is an arid region with no through-flowing streams, except the Colorado River to the south.

a. Study the Basin and Range province shown in Figure 8.1. What structure characterizes this province?

b. Study the map above and draw an idealized cross section of the area. Show the style and degree of deformation. How do the mountains in this area differ from those in the Landsat on page 204?

c. What active geologic and geophysical phenomena would you expect to find in this area?

d. Show, with arrows, the direction of the stresses that produced deformation in this area.

e. How is this area likely to evolve if the present tectonic forces continue?

f. Label the following features: (a) alluvial fans, (b) bajadas, and (c) playa lakes.

EXERCISE 19

DIVERGENT PLATE BOUNDARIES

OBJECTIVE

To become acquainted with the characteristics of divergent plate boundaries and to recognize the topographic expression of the geologic processes that occur as tectonic plates split and move apart.

MAIN CONCEPTS

Divergent plate boundaries are characterized by tensional stresses that typically produce long rift zones, normal faults, and basaltic volcanism. It is the tectonic setting where oceanic crust is generated.

SUPPORTING IDEAS

1. The oceanic ridge is a broad, fractured swell with a total worldwide length of about 70,000 km.
2. The oceanic ridge is broken into hundreds of segments by transform faults. Secondary segmentation occurs where ridge crests overlap.
3. The characteristics of a specific portion of the ridge depend on its spreading rate.
4. Patterns of magnetic reversals on the sea floor permit geologists to determine the age of the oceanic crust and rates of plate movement.
5. Divergent plate margins also develop within continents, causing continental rifting. Rift valleys and associated basaltic volcanism are typical surface expressions of continental rifting.

THE OCEANIC RIDGE

One of the most remarkable tectonic features on our planet is the oceanic ridge that encircles the entire globe and marks the boundary where tectonic plates split and move apart. As the plates move apart, the mantle beneath the rift becomes decompressed, allowing the hot mantle to partly liquefy and form a basaltic magma. This magma continuously erupts along the rift to form new oceanic crust. The oceanic ridge is thus the site of the most intense and continuous volcanism on Earth. This fact was not known until recently because the volcanic eruptions occurred beneath the ocean. Throughout Earth's long history, the geologic processes operating at divergent plate boundaries have been among the most fundamental processes that shape our world. Indeed, more rock is generated at divergent plate margins than by all other geologic processes combined. It is therefore important for you, as a student of geology, to become familiar with this unique region of our planet and to understand how it results from the Earth's internal energy.

CONTINENTAL RIFTS

Continental rifts are closely related to the oceanic ridge and are part of the ongoing tectonic activity of our planet.

They commonly produce striking physiographic features, intense volcanic activity, and seismicity. The study of continental rifts is important because of the opportunity to directly investigate rift processes in action. By their very nature, modern rifts are only a snapshot—an incomplete view of the rifting process. Ancient rifts, however, are numerous and have operated in a similar fashion for at least 2500 million years. Therefore, they have the potential to complete the picture of rift development. These include passive continental margins, such as the Atlantic coasts of the Americas, and aborted rifts, which provide a record of ancient plate tectonic activity not preserved in oceanic rifts.

PROBLEMS

This exercise introduces you to the landscape beneath the sea and to new information about divergent plate margins that we have learned from studies of both the ocean floor and rift valleys on continents.

1. Study the world map on pages 198–199, paying special attention to the mid-oceanic ridge. Trace its path around the globe noting how it changes from one area to another.

a. How does the "mountain belt" of the oceanic ridge differ from the mountain belts on the continents? Consider such features as aerial extent, width, height, configuration, and structural features expressed at the surface.

b. Compare the size (width and height) of the oceanic ridge in the Atlantic and in the east Pacific oceans. What evidence suggests that the ridge in the Pacific is spreading faster than that in the Atlantic?

c. Color the rift valley at the crest of the oceanic ridge red. Why is a rift valley better developed in the Atlantic and Indian ridges compared to that in the Pacific?

d. The oceanic ridge extends into continents in western North America, and in east Africa and the Red Sea. What structural and topographic features occur where the ridge extends into continental crust?

e. How many triple junctions on the oceanic ridge can you find on the global map? Why do they occur in the oceans and not on the continents?

f. How does this ridge terminate off the coast of northern California?

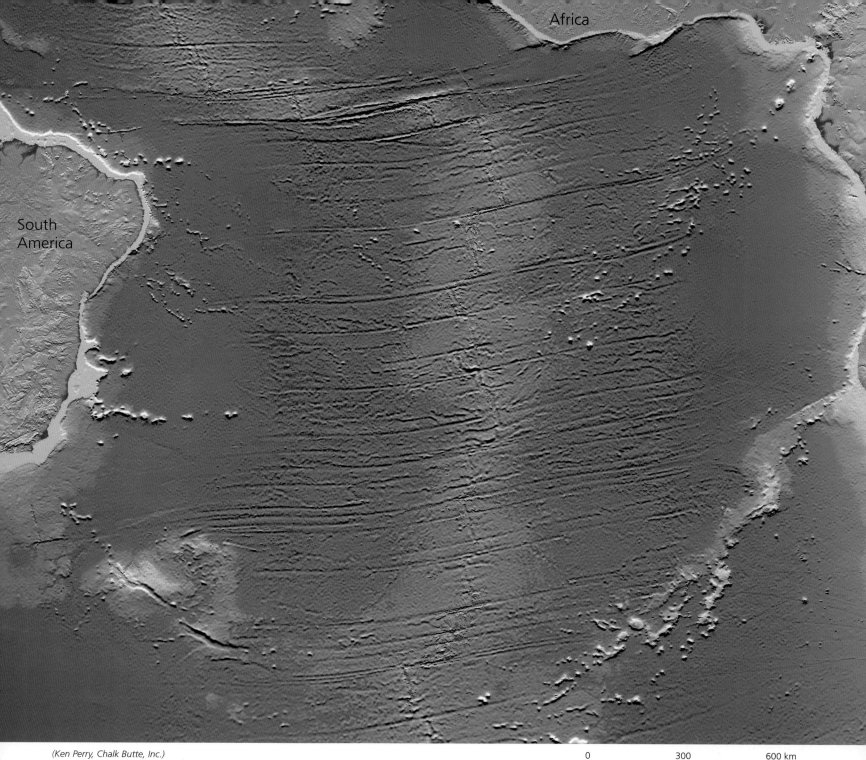

Africa

South
America

(Ken Perry, Chalk Butte, Inc.)

0   300   600 km

2. Mid-Atlantic Ridge

a. How do the structure and topography of the Mid-Atlantic Ridge support the theory of plate tectonics?

b. Note that the ridge is much wider in the south than in the north. Explain why such variations occur.

c. Color the rift valley red at the crest of the ridge. Work carefully on a piece of tracing paper at first and, when you are satisfied with your results, transfer your work to the map. Is the rift valley a continuous feature along the crest of the ridge or does it exist in only a few areas?

d. Note that the ocean floor is much deeper away from the ridge. Explain this in light of magmatic activity (heat) and density variations.

e. What is the origin of the steep slope (continental slope) that lies off the shore of South America and Africa?

f. Why does the trend of the ridge parallel the continental margins on both sides of the Atlantic Ocean?

South America

(D.T. Sandwell and W.H.F. Smith)

0   500   1000   1500 km

3. East Pacific Rise

a. How many tectonic plates are shown on this map?

b. Draw a series of arrows showing the direction of plate movement. What topographic and structural features indicate plate movement?

c. Study the crest of the oceanic ridge in this area. How does it differ from the crust of the Mid-Atlantic Ridge shown on page 207?

d. On the map, circle areas where the crest of the oceanic ridge overlaps.

e. Compare the size (width and height) of the East Pacific Rise with

the Mid-Atlantic Ridge (page 207). Explain what might cause these differences.

f. Locate the triple junctions of the plate boundaries in this area of the Pacific Ocean. What types of plate boundaries do they connect?

4. Crest of the Mid-Atlantic Ridge

This segment of the Mid-Atlantic Ridge crest was mapped in detail using the most sophisticated underwater mapping technology. The result was superposed on a smaller-scale map of lower resolution, in order to show the regional setting. Thus, the topography, in the areas beyond the ridge crest, appears blurred and indistinct. Study the ridge crest where topography details are clear and crisp. Mark the rift valley along the ridge crest with a highlighting pen. (Work on a piece of tracing paper at first and then transfer your results to the map. Then sketch the trace of the transform faults and fracture zones on the map.)

a. Note that elevation is shown by color tints—green is the lowest, red tones are intermediate, and gray-white is highest. Carefully sketch a cross section across the ridge near the middle of the map. Show the topographic profile and major faults.

b. Draw a series of small arrows showing the direction of movement of the plates away from the rift valley in each segment between the fracture zones.

c. How many transform faults offset the ridge?

d. What is the origin of the north–south trending linear ridges and troughs on either side of the rift valley?

e. What evidence indicates that the offset along the transform faults is largely strike-slip movement?

(Lamont Doherty Observatory)

| 0 | 100 | 200 km |

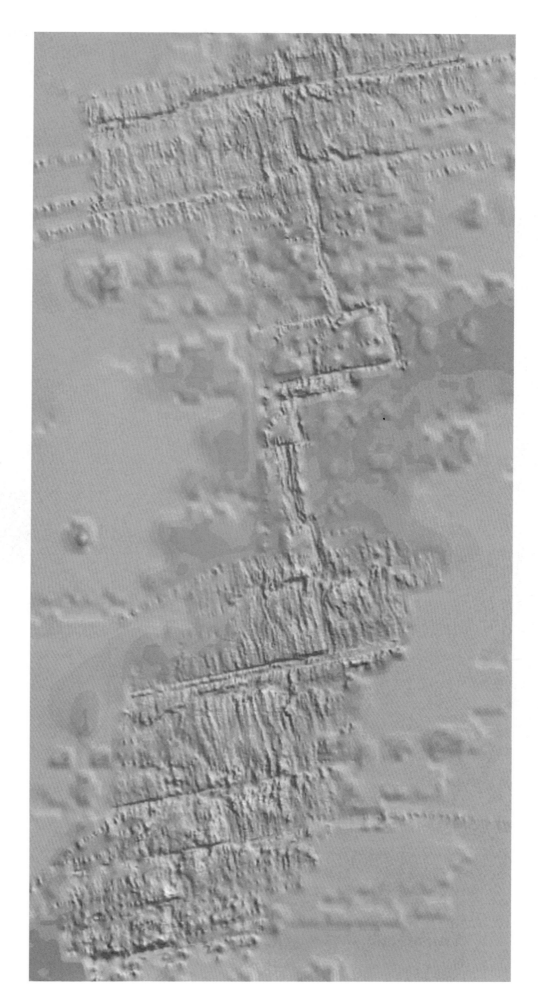

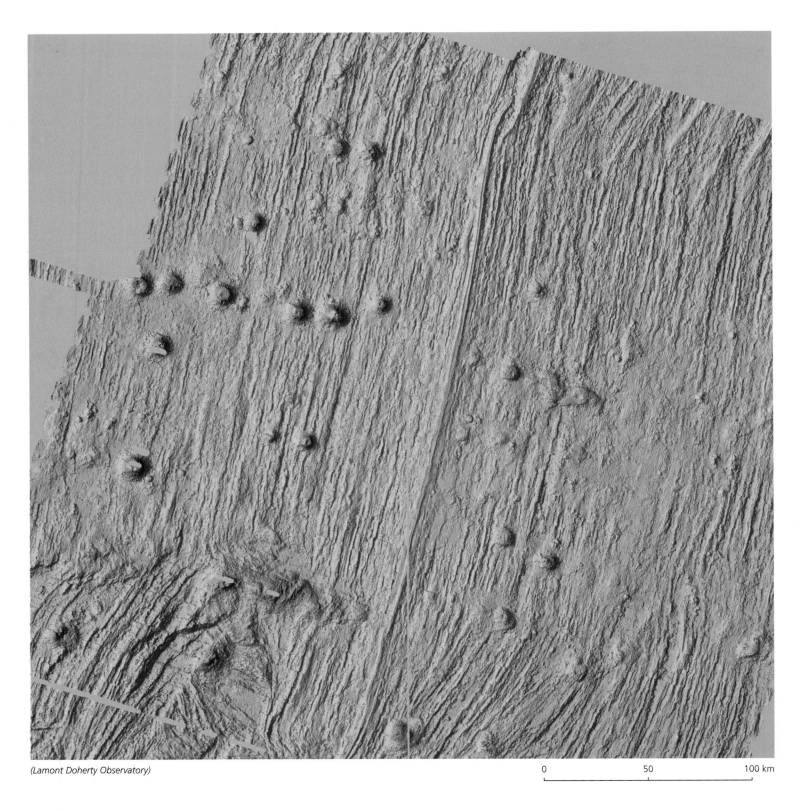

(Lamont Doherty Observatory)

0   50   100 km

5. Crest of the East Pacific Rise

This image, like that shown on page 209, shows the remarkable details of the topography of the oceanic ridge crest and the features produced at divergent plate margins.

a. Explain the origin of the circular hills scattered throughout the area.

b. What structure and topographic features mark the crest of the ridge?

c. How does the crest of the East Pacific Rise compare with that of the Mid-Atlantic Ridge (page 209)?

d. Explain the origin of the elongate ridges and troughs that parallel the ridge crest.

e. Sketch a cross section across the ridge and ridge flanks near the center of the image. How does this profile compare with the profile made for the Mid-Atlantic Ridge?

f. Circle the area where the crest of the ridge overlaps.

6. Ridge Segment Adjacent to a Fracture Zone

The intersection of the ridge crest and the transform fault (marked by a deep valley near the top of the image) is a zone of complicated tectonic and magmatic activity. The ridge that marks the spreading axis of the plates terminates abruptly at the transform fault and is offset to the east.

a. How do we know the ridge is displaced to the east and not to the west?

b. Explain why the ridge crest appears to extend north across the fracture zone for a short distance.

c. What causes the numerous linear features that trend north–south across the entire area? How is this fabric related to patterns of magnetic reversals on the sea floor?

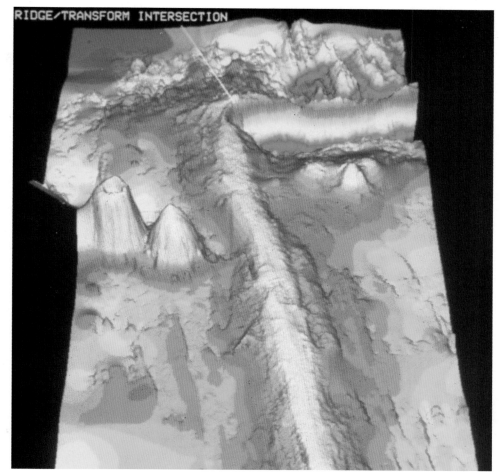

RIDGE/TRANSFORM INTERSECTION

(K.C. Macdonald)

7. Juan de Fuca Ridge

This area shows a small segment of the Juan de Fuca Ridge off the coast of Washington State. It reveals many details of the tectonic and igneous processes that occur along divergent plate margins.

a. What are the major structural features in this area? How do they differ from the structures in transform fault systems?

b. Highlight the volcanoes in this area with a color felt-tip marker. What types of volcanoes are they?

c. What is the diameter of the largest volcano? How does it compare with the diameter of Mount St. Helens?

d. Lava flows are characterized by irregular lobate margins such as those shown on page 69. Carefully study the surface features of this area. Map the margins of probable lava flows. How are the lava flows associated with the major faults?

e. Make a map of the faults in this area. What types of faults are they?

(Lamont Doherty Observatory)

0   10 km

8. Paleomagnetism—Atlantic Ocean

When a lava cools and solidifies, the magnetic domains within the iron minerals contained in the lava align in the direction of Earth's magnetic field. The solidified lava thus preserves a record of Earth's magnetic field at the time the rocks were formed. As early as 1906 scientists recognized that the poles of the magnetic field preserved in some rocks were oriented in the opposite direction from the poles evident in other specimens, as if Earth's north and south magnetic poles had switched places.

Subsequent studies showed that these magnetic reversals do, indeed, occur, and that the magnetic poles have been reversed many times during the geologic past. The effect of these reversals has been to create periods of "normal" magnetism (periods in which the polarity matched the present position of the north and south magnetic poles), and periods when the magnetic field was reversed.

To test the plate tectonics theory, scientists proposed that, if new oceanic crust is created at the oceanic ridges by the extrusion of lava, then the rocks on the ocean floor adjacent to the ridges should preserve a record of the periods of normal and reversed magnetism. If we measure the strength of Earth's magnetic field across the ridge, a series of anomalies, or differences, should occur. Where rocks solidified during periods of "normal" magnetism, higher values should be found because induced magnetism adds to the present strength of the magnetic field. Lower values should be found in rocks solidified during reversed magnetism, because the polarity of these rocks reduces the local effect of Earth's present magnetic field.

If we were to map the magnetic anomalies across the oceanic ridge, alternating bands of high and low magnetism should appear on both sides of the ridge, one side being the mirror image of the other. Such a pattern would provide strong evidence for sea-floor spreading and the theory of plate tectonics.

The map on the right is of a portion of the Atlantic Ocean and the magnetic measurements made by a research vessel as it crossed the ridge on four traverses. When the curve on a traverse is above zero, the strength of the magnetic field is greater than normal (the paleomagnetism in the rocks is adding to the strength of the present magnetic field). Where the curve drops below zero, the strength of the magnetic field is less than normal because the paleomagnetism in the rocks has a reverse polarity and thus reduces the present strength of the magnetic field.

The objective of this exercise is to determine the zones of normal and reverse polarity along each track of the research vessel and to correlate the zones along this part of the Mid-Atlantic Ridge.

a. On curve A, note the points where the magnetic curve intersects the line of zero field strength. To emphasize the negative anomalies (areas below the zero line of the curve), shade them gray with a soft pencil. The first two adjacent to the ridge crest are done to illustrate. Do the same for curves B, C, and D.

b. On curve A, color the positive magnetic anomalies as follows: youngest = red, next youngest = orange, next youngest = light green, and next youngest = dark green. Do the same for curves B, C, and D.

c. The Mid-Atlantic Ridge is offset by numerous fracture zones. We have generalized this segmentation of the ridge by plotting only a few major fracture zones between each traverse shown on the map by a dark heavy line. The next step is to interpolate the patterns of magnetic anomalies between each traverse. This can be done by projecting the crest of the ridge perpendicular to the fracture zone and then extending each positive anomaly from the measured curve up to the fracture zone. This creates bands of magnetic anomalies offset by fracture zones.

d. Color all positive anomalies red, orange, light green, and dark green, as you did on the measured curves.

e. Describe, briefly, the patterns of the magnetic anomalies shown on the map that you have constructed. Do the patterns cross or parallel the oceanic ridge? Are the patterns on either side of the ridge similar or different?

f. Write a brief paragraph explaining how the pattern of magnetic anomalies on the sea floor supports the theory of plate tectonics.

g. Refer to the geomagnetic time scale in your textbook and determine the age of each band on the map.

h. Calculate the average rate of spreading of the floor of the Atlantic Ocean during the last 3 million years. (Use the map scale and the time duration of the rock units determined from the geomagnetic time scale in your text.)

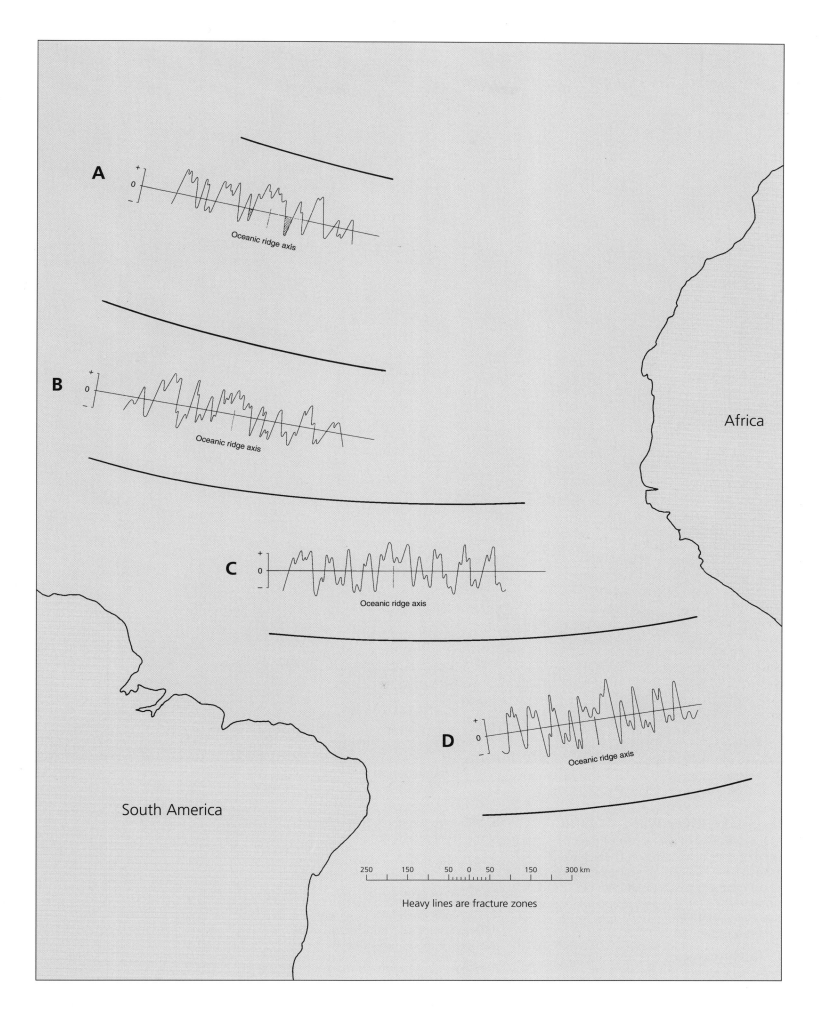

A

Oceanic ridge axis

B

Oceanic ridge axis

Africa

C

Oceanic ridge axis

D

Oceanic ridge axis

South America

250   150   50 0 50   150   300 km

Heavy lines are fracture zones

(U.S. Department of Agriculture)

Color infrared photo

0   0.5 miles

9. Snake River Plain—Idaho

The Snake River Plain is part of the Columbia Plateau, a region characterized by extensive floods of basalt. The region is very flat and has been the site of numerous recent volcanic eruptions associated with rifting in western North America.

   a. What type of volcanic eruption characterizes this area?

b. What are the linear features east of the most recent flows?

c. Why are there no volcanic cones in an area so dominated by the extrusion of lava?

d. Highlight each lava flow with a different color and indicate the relative age of the various flows.

10. Rio Grande Valley, New Mexico

The Rio Grande rift valley extends from Mexico northward into Colorado.

a. Map the faults that mark the boundary of the rift system.

b. Map the major volcanoes in this area.

c. What type of volcanism is associated with the rift?

d. Draw a generalized cross section across the rift in the central part of the map.

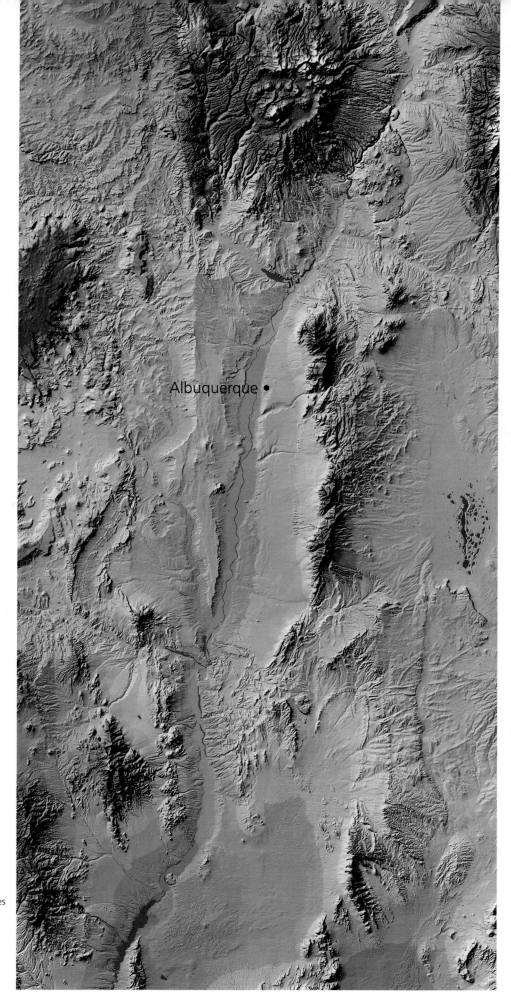

Digital shaded relief map
(*U.S. Geological Survey*)

0   30 miles

11. East African Rift Valleys

a. Study the topography of eastern Africa and draw lines along the boundaries of the rift valleys. (Work on a piece of tracing paper at first and then transfer your results to the map.)

b. How do rift valleys differ from stream valleys and glacial valleys? Explain the reason for this difference.

c. How does the submarine topography of the Red Sea compare with the rift valleys?

d. How much of the Red Sea is underlain with oceanic crust?

e. Is there any evidence that volcanism is associated with the rift system?

f. Make a sketch map showing how Africa will appear a few million years from now if present tectonic processes continue.

g. What evidence indicates that the African rift valleys are connected to the divergent plate boundaries of the Indian Ocean and the Red Sea?

h. Why does East Africa have such a high elevation?

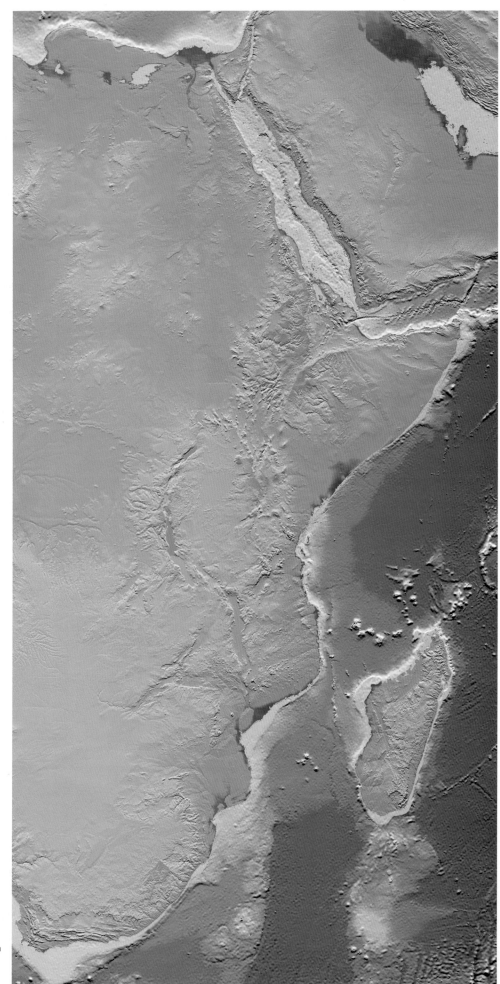

(Ken Perry, Chalk Butte, Inc.)

0	1000	2000 km

12. Atlantic Margins

The margin of the eastern South American continent is not the shoreline but the edge of the continental shelf (shown in lightest blue tones).

a. Describe the trend of the continental margin. How does it compare with that of the rift valleys of eastern Africa?

b. Why is the continental margin so steep in most areas?

c. Note that the Amazon River is depositing most of its sediment load as a delta on the continental shelf and as a deep-sea fan. What types of rock probably underlie the Amazon sediments?

d. What geologic process will dominate along the continental margins of South America in the next few million years?

e. What will be the ultimate fate of the Amazon sedimentary sequence assuming future shifts and changes in plate movement?

(Ken Perry, Chalk Butte, Inc.)

0   1000 km

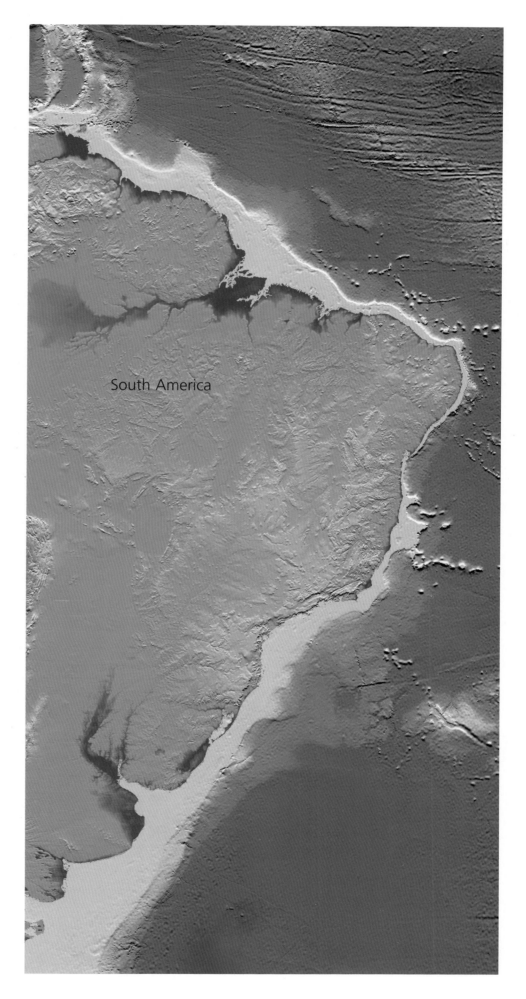

South America

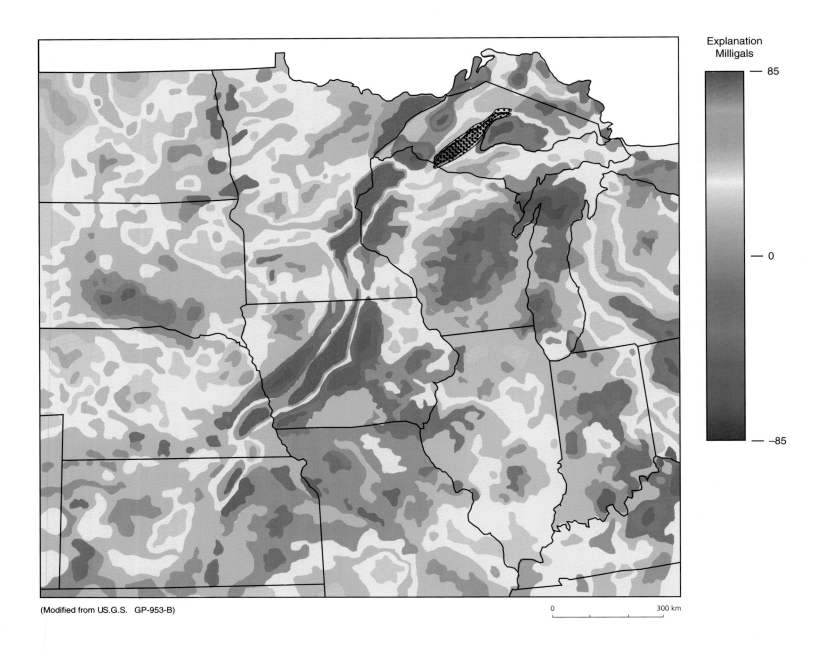

(Modified from US.G.S. GP-953-B)

0   300 km

Explanation
Milligals

— 85

— 0

— −85

13. Mid-Continent Gravity Anomaly

This map shows variations in the gravity field in the central part of the United States. Exceptionally high gravitational field is shown in red, yellow shows normal gravity, and blue shows a low gravity field.

a. What is the length and width of the mid-continent gravity high? How does the size of this feature compare with the East African rift valleys and the Rio Grande Rift?

b. The rock sequence along the Keweenawa Peninsula of northern Michigan consists of a sequence of basalt and conglomerate nearly 80,000 feet thick. These are shown with a stipple pattern. How are these rocks related to the mid-continent gravity anomaly?

c. The density of common rock is as follows: granite = 2.6, basalt = 3.0, sedimentary rocks = 2.5, gneiss and schist = 2.6. What rock type likely

exists in the mid-continent gravity high in Kansas, Iowa, and Minnesota?

d. Draw a cross section across the mid-continent gravity high showing the structure and rock types that could produce this gravity anomaly.

TRANSFORM PLATE BOUNDARIES

OBJECTIVE

To become acquainted with the characteristics of transform faults and their associated fracture zones and to recognize the role they play in the global tectonic system.

MAIN CONCEPT

Transform plate boundaries are shearing zones where tectonic plates grind and slide horizontally past each other. They are intimately related to other types of plate boundaries in that they provide a mechanism by which plate segments are connected.

SUPPORTING IDEAS

1. Transform faults in the ocean basins are closely associated with the oceanic ridge.
2. Oceanic fracture zones are prominent linear features extending beyond the active transform faults.
3. There are three major types of transform faults: (a) ridge–ridge, (b) ridge–trench, and (c) trench–trench.
4. In the shearing processes involved with transform faults, secondary features are created, such as parallel ridges and valleys, pull-apart basins, and belts of secondary folds.
5. Transform faults juxtapose cold oceanic crust against a hot ridge axis.
6. Continental transform faults are similar to oceanic transforms but lack the fracture zone extensions.

CHARACTERISTICS OF TRANSFORM PLATE BOUNDARIES

Transform plate boundaries are strike-slip faults, along which tectonic plates slide horizontally past each other. The trend of the transform is commonly parallel to plate movement, and the fault plane is essentially vertical, cutting through the entire lithosphere. Transform faults occur along the entire oceanic ridge and also cut across continents connecting various types of plate boundaries. Thus, they are a worldwide tectonic feature.

Transform faults typically produce shallow earthquakes as the plates move past each other, but only minor volcanic activity occurs along this type of plate boundary.

Transform faults are a special type of fault that *transforms* one type of plate motion to another. For example, the diverging motion between plates between adjacent segments of the oceanic ridge may be changed (transformed) from one direction to another, or the diverging motion between plates at a ridge may be transformed into converging motion between plates at a subduction zone. There are three major types of transform plate boundaries, (1) ridge–ridge, (2) ridge–trench, and (3) trench–trench.

Ridge–ridge transform faults are by far the most abundant. They occur at roughly 100 km intervals throughout the entire ridge system and, together with their associated fracture zones, form one of the most remarkable structural fabrics on our planet. It is important to remember that active ridge–ridge transform faults occur *only* between ridge segments. However, their associated fracture zones may extend up to 10,000 km long.

The structure and topography of fracture zones depend largely on two things: (1) the temperature and, therefore, age difference between the rocks on either side of the fracture zone, and (2) the spreading rate of the oceanic ridge. Younger oceanic crust is hotter than older crust and is thus expanded and thickened. As a result, where young hot crust is displaced against older cold crust, a cliff forms on the side of the fracture with the hot expanded younger rock.

Slow-shearing transforms produce narrow, well-defined valleys with steep walls. Fast shearing produces a broader zone of deformation, up to 100 km wide, in which small ridge segments may be incorporated into the fault zone.

Ridge–trench transforms are much less common but similar in general form to ridge–ridge transforms. They are commonly more rugged and irregular and are not associated with fracture zones.

Trench–trench transforms are also rare but form important connections where two subduction zones dip toward each other.

Continental transform faults are similar to those on the ocean floor, but they cut through thicker and more complex continental lithosphere, and their topographical expression is considerably modified by stream erosion. They are much less common, because transform faults are associated with divergent plate margins that, when developed on continents, soon evolve into new ocean basins.

PROBLEMS

1. Study the world map on pages 198–199 and locate two examples of the following types of transform faults:

    a. Ridge to ridge

    b. Ridge to trench

    c. Trench to trench

    d. Trench to mountain range

    e. Ocean ridge to mountain range

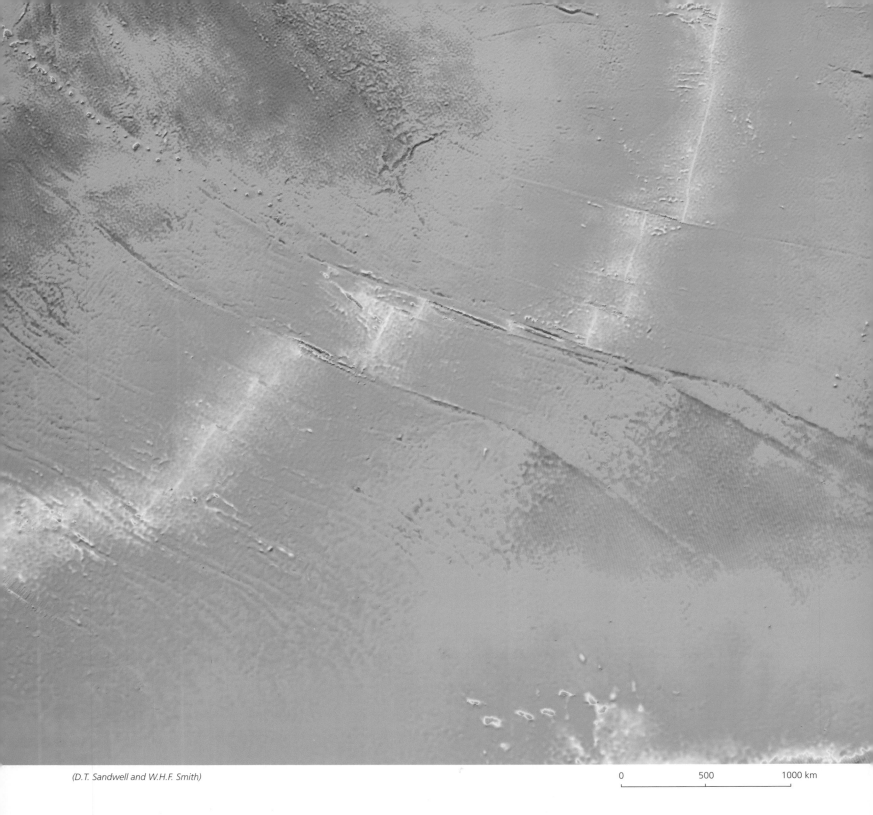

(D.T. Sandwell and W.H.F. Smith)

0   500   1000 km

2. East Pacific Rise—Southeast Pacific

a. Locate the longest transform faults in this area. What is the distance of maximum offset of the ridge?

b. How long are the associated fracture zones?

c. What is the average distance between fracture zones?

d. Study the details of the larger fracture zones in this area. Note the depressions or valleys along the fracture zone and the boundary scarps or cliffs that form prominent walls adjacent to the narrow valley floors. Are the valleys simple troughs bounded on both sides by cliffs?

e. Sketch a series of topographic profiles across the fracture zone east and west of the oceanic ridge

to show variations in topography along the valley walls.

f. What direction does the fault scarp face in the segments west of the ridge? What direction does the fault scarp face in the segment east of the ridge? Explain why these changes occur.

g. What evidence indicates that the fault plains are nearly vertical?

3. Mid-Atlantic Ridge

a. How many well-defined *fracture zones* occur in this area? What is their average length?

b. What is the average length (offset) of the *transform faults* in this area?

c. How do the fracture zones indicate the path of the spreading continents?

d. How do the depth and width of the fracture zones compare with the depth and width of the rift valley on the oceanic ridge?

e. Do the junctions between transform faults and spreading ridges occur at right angles or at various angles?

f. How does the orientation of transform faults indicate plate movement?

g. Why are earthquakes limited to the transform section of the fracture zone?

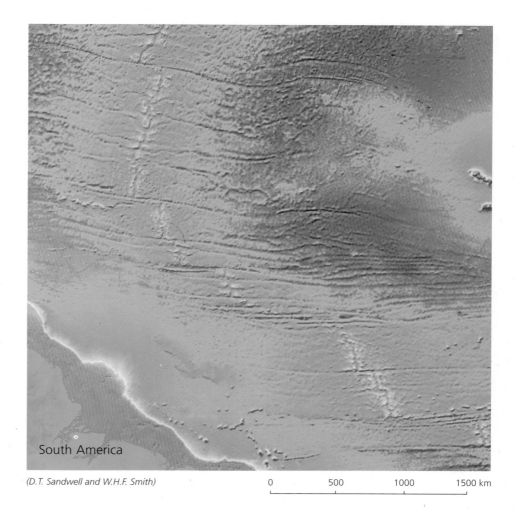

South America

(D.T. Sandwell and W.H.F. Smith)

0	500	1000	1500 km

4. Details of Transform Fault along the Mid-Atlantic Ridge

a. Outline the crest of the Mid-Atlantic Ridge with a highlighting pen and draw the trace of the transform faults. Is the transform fault a single fracture or a complex of several fault planes?

b. What is the origin of the east–west trending ridge in the center of the transform fault zone?

c. Is there evidence of minor extrusions associated with this transform system?

d. How far has the oceanic ridge been displaced along the transform fault?

e. What is the deep north–south valley (expressed in blue) that cuts across the axial ridge in the transform fault zone?

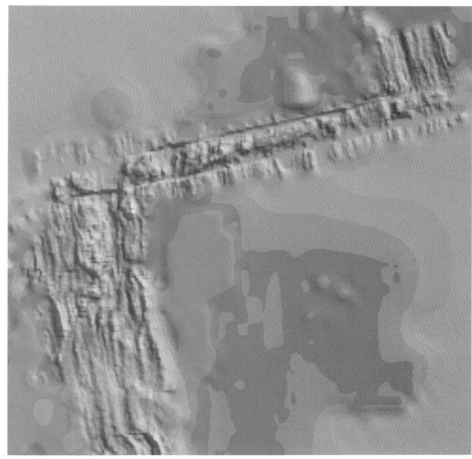

(Lamont Doherty Observatory)

0	100	200 km

221

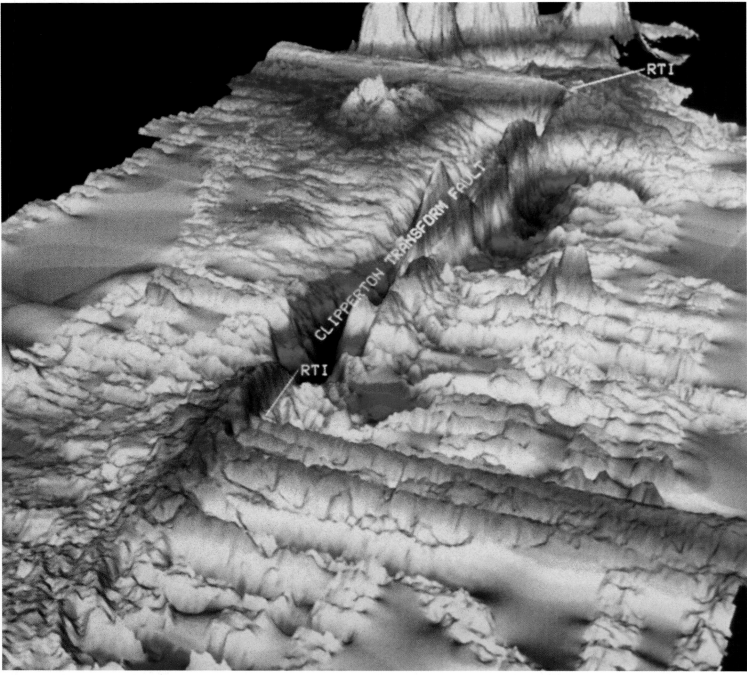

(K.C. Macdonald)

5. East Pacific Rise

a. How does the hot ridge axis affect the adjacent cold lithosphere at the junction of the oceanic ridge and the transform fault?

b. How wide is the transform fault and fracture zone (compared to the rift valley of the mid-oceanic ridge)?

c. What structural and topographic features occur within this fracture zone?

d. Is there evidence that minor volcanism is associated with the fracture zone?

e. What is the origin of the numerous linear ridges that parallel the ridge and extend across most of the region?

6. New Zealand

a. Show with a series of arrows the direction of plate movement.

b. What is the inclination of the subduction zone north of New Zealand (Tonga Trench) and south of New Zealand (Macquarie Trench)?

c. Trace the Alpine fault that extends across New Zealand and connects these trenches. Why is the Alpine fault considered to be a transform fault?

d. Draw the transform fault extending southeastward from the southern part of the Macquarie Trench. Show arrows indicating relative movement. What type of transform fault is this?

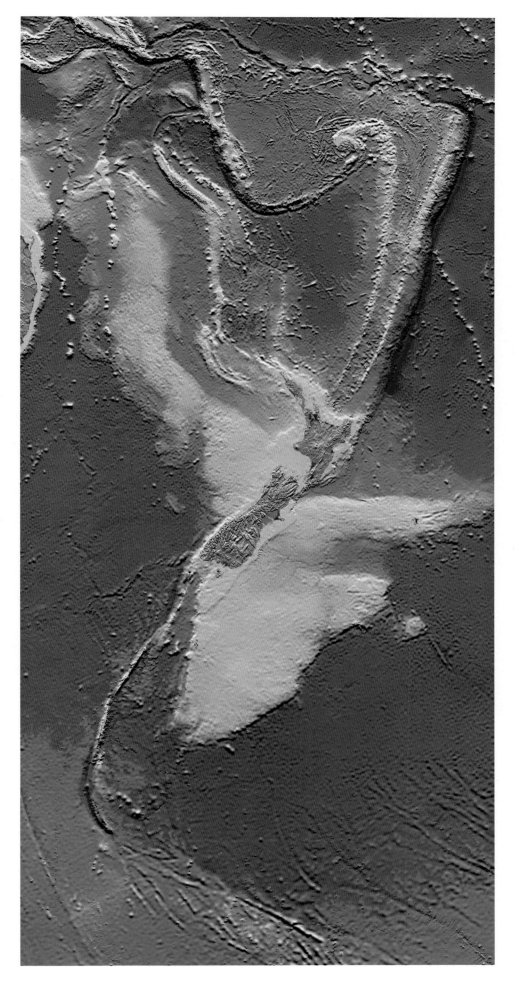

(Ken Perry, Chalk Butte, Inc.)

0	500	1000	1500 km

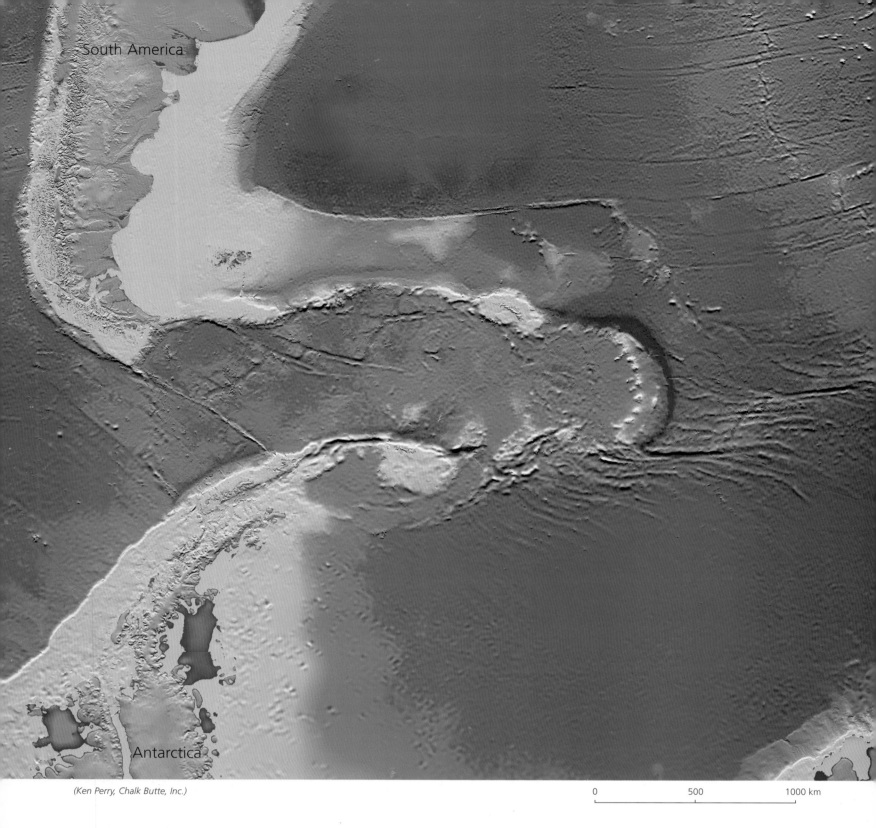

South America

Antarctica

(Ken Perry, Chalk Butte, Inc.)

0   500   1000 km

7. Scotia Arc

a. Study the tectonic elements shown in this area and label the following features: (1) transform faults, (2) trench, (3) folded mountain belt, (4) rift valley of oceanic ridges, and (5) magmatic arc.

b. What types of transform faults are found in this area?

c. What is the origin of the spreading ridge between South America and Antarctica?

d. What is the relationship between the mountain belt in Antarctica and the Andes Mountains of South America?

8. The Dead Sea Transform

The Dead Sea transform fault system connects the divergent plate boundary of the Red Sea with the convergent plate boundary between the Arabian and Turkish plates. It is marked by the depressions of the Gulf of Aqaba, the Dead Sea, the Sea of Galilee, and the Beka Valley. This is a satellite image of part of the Dead Sea transform system.

a. Draw a red line along the faults that make up the Dead Sea transform system (the faults are generally expressed by steep cliffs).

b. Show the direction of displacement with a series of arrows.

c. What is the origin of the Dead Sea depression?

d. What is the origin of the Palmyra Hills near the top of the image?

e. Draw a simple map showing the faults of the Dead Sea transform system and the depressions and hills of parts b and c. Show with arrows the stress conditions that exist in this area.

(Earth Satellite Corporation)

```
0        20        40 km
|---------|---------|
```

Salinas River

Great Valley

San Luis Obispo •

(U.S. Geological Survey)   Digital shaded relief map

9. San Andreas Fault

a. Study the landscape in this area and draw lines (map) on the major faults. What evidence indicates that these are strike-slip faults and not thrust or normal faults?

b. Refer to the block diagram on page 171 and label the following features on the map: (a) slivers, (b) elongate troughs, (c) elongate ridges, and (d) offset drainage.

c. Show the direction of relative movement along the faults with arrows.

d. Points A and B are parts of the same formation which formed 10 million years ago. What has been the total displacement during the last million years? What has been the average rate of displacement (cm/year)?

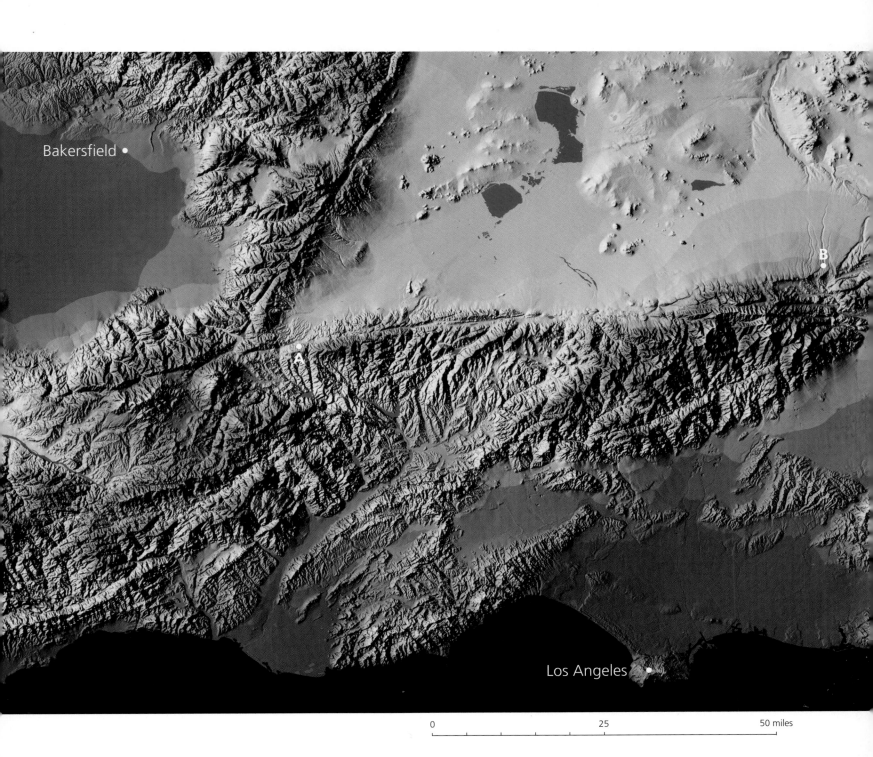

Bakersfield

B

A

Los Angeles

0   25   50 miles

CONVERGENT PLATE BOUNDARIES

■ OBJECTIVE

To become familiar with the major types of convergent plate boundaries and the structure and landforms they produce.

■ MAIN CONCEPT

Convergent plate boundaries are zones where lithospheric plates collide and one descends down into the mantle. They are zones of intense structural deformation, mountain building, magmatic activity, and metamorphism.

■ SUPPORTING IDEAS

1. There are three types of convergent interactions: (1) convergence of two oceanic plates, (2) convergence of an oceanic and a continental plate, and (3) convergence of two continental plates.
2. Most subduction zones are marked by (1) an outer swell, (2) a trench, (3) a magmatic arc, and (4) a backarc basin.
3. Silicic magma is generated at a subduction zone by partial melting of the descending lithosphere.
4. Crustal deformation at subduction zones produces a melange in the forearc and extension or compression in the backarc areas.
5. Continental collision always produces strong compression that causes folding, thrust faulting, and the formation of a mountain range.

CHARACTERISTICS OF CONVERGENT PLATE BOUNDARIES

Field studies of mountain belts and island arcs, combined with geophysical measurements and oceanographic surveys, give us an understanding of the fundamental characteristics of convergent plate boundaries and the geologic processes that occur along them. Three distinct types of convergence are recognized: (1) ocean to ocean, (2) ocean to continent, and (3) continent to continent.

Ocean-to-Ocean Convergence

Where two oceanic plates collide, one is thrust under the other, forming a subduction zone. The descending plate is heated, triggering the generation of magma, which is buoyant (less dense than the surrounding rock). The magma rises to the surface and erupts to build an island arc. Several important structural and topographic features commonly form at many subduction zones: (1) a broad rise or bulge, known as an *outer swell,* develops just before the descending plate dives down into the mantle, (2) a deep sea *trench,* (3) an *accretionary wedge* made of highly

deformed sedimentary and metamorphic rocks, (4) a *magmatic arc,* and (5) a backarc basin.

Ocean-to-Continent Convergence

Where an oceanic plate collides with a continental plate, the less-dense continental plate always resists subduction and overrides the denser, descending oceanic plate. An outer swell and a deep sea trench develop on the side of the descending plate. On the overriding plate, the continental margin is deformed into a folded mountain belt and magma erupts to form a chain of andesitic to rhyolitic volcanoes with subordinate basalt.

In addition to the topographic and structural features formed in ocean-to-ocean and ocean-to-continent convergence, earthquakes occur in a zone inclined downward beneath the adjacent island arc or continent and outline the surface of the descending plate. Earthquakes as deep as 600 km may occur.

Continent-to-Continent Convergence

Where two continental plates collide, both resist subduction because conti-

nental crust is too buoyant to descend down into the mantle. Therefore, one plate tends to override the other for some distance to form an exceptionally thick mass of continental crust. Consequently, there is no outer swell, trench, or forearc wedge. Instead, the continental crust of the two plates is fused together, commonly trapping segments of oceanic crust between them.

PROBLEMS

This exercise involves analysis of the world map on pages 198–199, and detailed maps and satellite images of important convergent plate margins. Remember, these maps are extremely accurate scale models of Earth's surface, made from the most recent elevation data sets obtained from various government agencies. Serious study of these models is the best way to gain an accurate understanding of the dynamics of our planet.

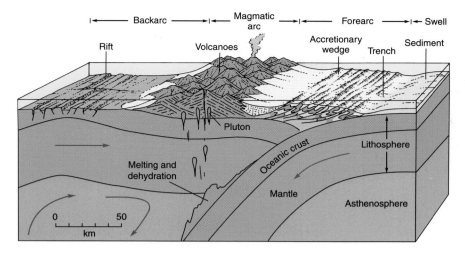

FIGURE 21.1
Ocean-to-Ocean Convergence

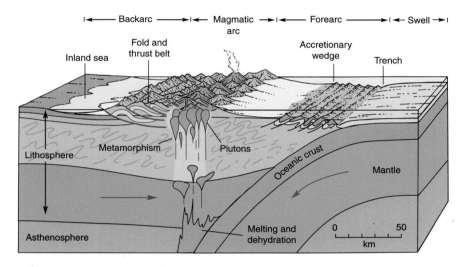

FIGURE 21.2
Ocean-to-Continent Convergence

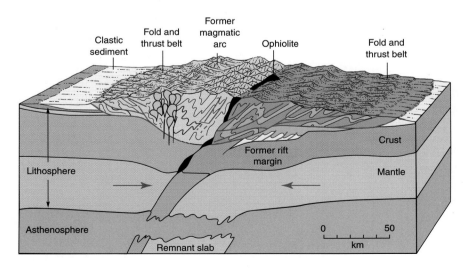

FIGURE 21.3
Continent-to-Continent Convergence

1. Global Map (pages 198–199)

a. With a highlighter pen, draw the zones (both active and inactive) of continent-to-continent convergence.

b. Are the mountains of eastern Australia tectonically active? What is the evidence for your conclusion?

c. Where is the active mountain belt on the Australian plate?

d. What geologic features occur where the following folded mountain belts extend from the continents to ocean basins?

• "Rocky Mountains" in southern Alaska extending westward

• Andes Mountains in southern South America

• Himalayan Mountains in southeast Asia

• Folded mountain belt of the Antarctic Peninsula

Explain the overall pattern you have discovered.

e. If both the Ural Mountains in Russia and the Himalaya Mountains are the result of continent–continent convergence, why are the Urals so low and the Himalayas so high?

f. From the standpoint of tectonic setting, how might the Mediterranean Sea and the Black Sea be considered similar to the Bering Sea and the Sea of Japan?

g. In light of present tectonic processes, which would be the younger mountain range?

• In Alaska: The Alaskan Range to the south or the Brooks Range to the north?

• In North America: The Rockies of western North America or the Appalachian Mountains of the eastern United States?

• In Australia: The Great Dividing Range on the east coast or the mountains of New Guinea?

2. Convergent Boundaries—West Pacific

a. What are the relatively high areas immediately east of the major trenches? Explain their origin.

b. Draw arrows to show the direction of plate movement in the Tonga trench area just north of New Zealand. Note that there are very few islands associated with it. What evidence indicates significant volcanism is associated with this subduction zone of the Tonga trench? What is the origin of the two ridges that parallel the trench?

c. What is the origin of the small, deep ocean basins that lie between the island arcs of Japan and the Philippines and the continental landmass of eastern Asia?

d. What evidence suggests backarc spreading occurred in the South China Sea?

e. What type of rock body will the numerous islands of the Pacific form when plate motion eventually transports the islands to the subduction zone.

f. If present plate motion continues, where might the Emperor Seamounts form an accreted terrain?

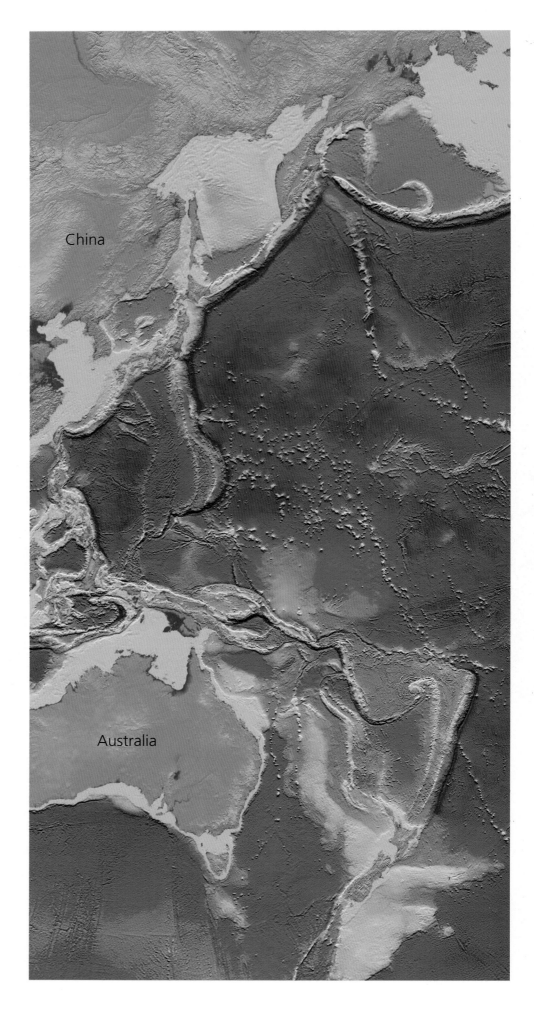

(Ken Perry, Chalk Butte, Inc.)

| 0 | 1000 | 2000 | 3000 km |

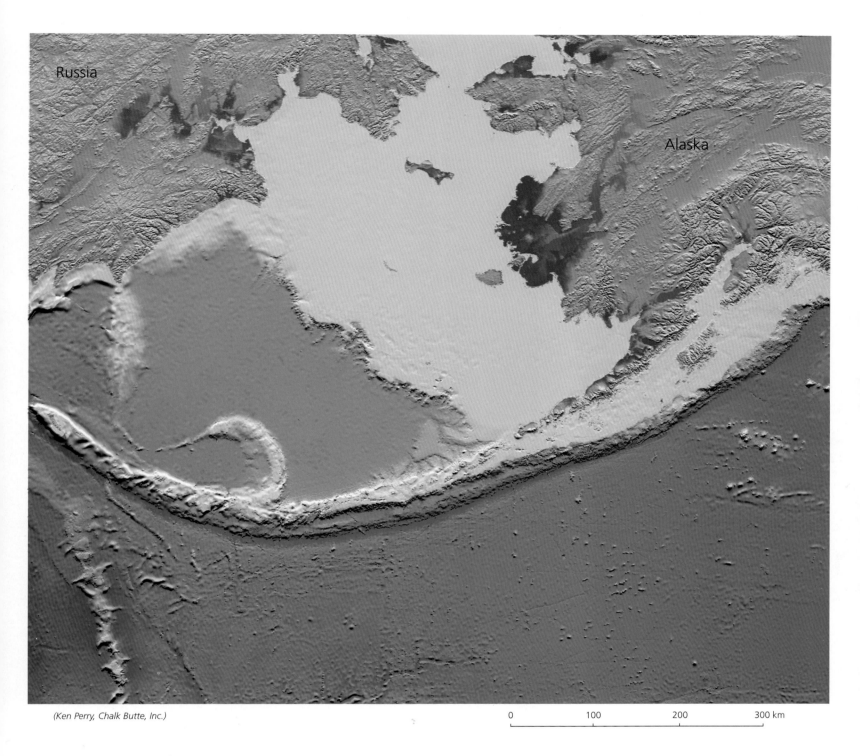

(Ken Perry, Chalk Butte, Inc.)

| 0 | 100 | 200 | 300 km |

3. The Aleutian Islands

a. With a series of arrows, show the direction of plate motion in this area. Which slab is being subducted?

b. Label the following features on the map:

(1) Outer swell

(2) Trench

(3) Accretionary wedge

(4) Magmatic arc

(5) Backarc basin

c. Is the Aleutian trench a zone of ocean-to-ocean convergence, ocean-to-continent convergence, or both? Explain the reasons for your answer.

d. How is the Aleutian trench related to (1) the Rocky Mountains, and (2) the Japan-Kuril trench?

e. What is the origin of the deep basin of the Bering Sea?

4. South America

a. Show with a series of arrows the direction of plate movement in this area.

b. What types of rock sequences would you expect to find in the central Andes Mountains?

c. How does the southern part of Central America differ in structure and rock types from the central Andes?

d. Judging from the topography, where would you expect the highest rate of plate movement to occur (southern, central, or northern Andes)?

e. Draw a cross section across the central Andes and the adjacent deep sea trench.

f. Where would deep focus earthquakes occur in this area?

g. Why are earthquakes and volcanoes so common in the Andes Mountains and are absent in the Appalachian Mountains?

h. Compare and contrast the converging plate margins in (1) the Aleutian Islands, (2) the Andes Mountains, and (3) the Himalayan Mountains. Write a brief paragraph explaining the differences in these areas.

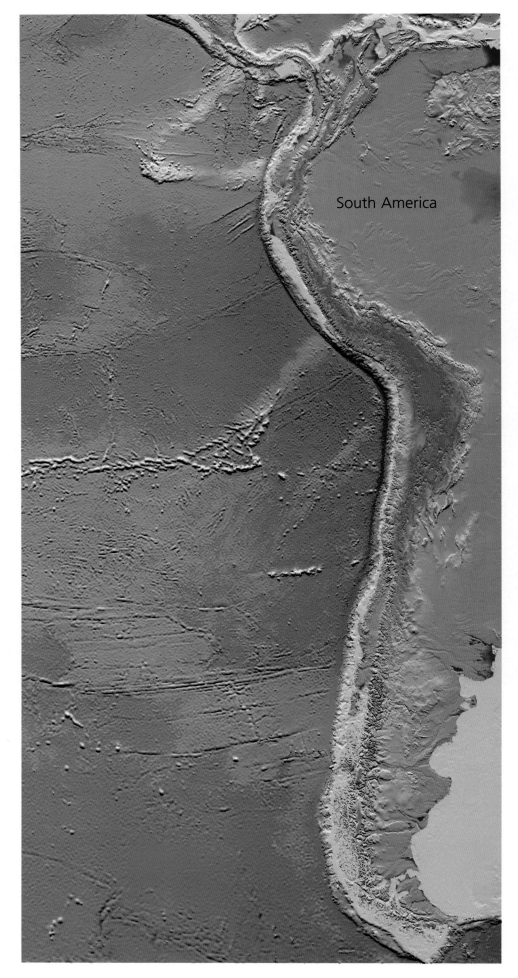

South America

(Ken Perry, Chalk Butte, Inc.)

| 0 | 500 | 1000 km |

Ancient
Seafloor

BEAUFORT SEA

Continental Fragment

Island Arc

Continental Fragment

Isl. Arc

Continental Fragment

Isl. Arc

ALASKA

Ancient Seafloor–Perm.

Stable
Platform

Ancient Seafloor – Trias.-Cretaceous

Deep Marine

Deformed
Margins of
N.A. Continent

Gre

GULF OF ALASKA

Isl. Arc

Continental Fragment

Stable
Platform

(U.S. Geological Survey)

Digital shaded relief map

| 0 | 100 | 200 | 300 | 400 miles |

5. Accreted Terrains—Alaska

Accreted terrains are an important component of most continents and provide important information about the movement of tectonic plates, and the processes occurring at convergent plate margins. Each terrain is bounded by major faults and is distinguished from its neighboring terrains by rock types, age, fossils, magnetic properties, etc. This map shows the general boundaries between the major terrains.

a. How does the topography of Alaska differ from that of the Appalachian Mountains (page 204)?

b. Review the stratigraphic sequence and structure shown on the geologic maps of the Grand Canyon (page 178), Michigan (page 180), and Wyoming (page 182). How do the rock types and structure of the accreted terrains in Alaska differ from those areas?

c. Why do the rock types and structure of Alaska suggest an accreted terrain rather than "normal" continental geology?

d. Would you consider the faults separating the major terrains to be thrust faults, normal faults, strike-slip faults, or lines of suture? Cite evidence for your answer.

233

6. Southern Cascade Mountains

a. Study the map on pages 198–199. What is the origin of the Cascade Mountains?

b. Why do the Cascade Mountains terminate abruptly in northern California?

c. Why does the volcanic field terminate in southern British Columbia?

d. If there is a subduction zone associated with the Cascade volcanoes, why isn't there a trench off the coast of Washington and Oregon?

e. How many volcanoes can you identify in this area?

f. What is the spacing between the volcanoes?

g. What types of volcanoes form the higher peaks?

h. What types of volcanoes form the smaller cones?

i. Why is the area west of the Cascades so highly dissected compared to the adjacent volcanic field to the east?

j. Identify the more recent lava flows on Mt. Shasta and emphasize them with a highlighter pen or colored pencil. What type of lava formed these flows? Cite evidence to support your answer.

k. Which of the composite volcanoes is the oldest? Cite evidence for your answer.

l. What types of volcanic features form the small hills scattered throughout the area?

m. Identify several structures that might be silicic domes.

n. Consider the nature of the geologic provinces adjacent to the Cascade Mountains. What might be the origin of the flat surface upon which the volcanoes were constructed?

o. Explain the origin of the low linear cliffs trending N–S in the central part of this area.

Digital shaded relief map

(U.S. Geological Survey)

0   25   50 km

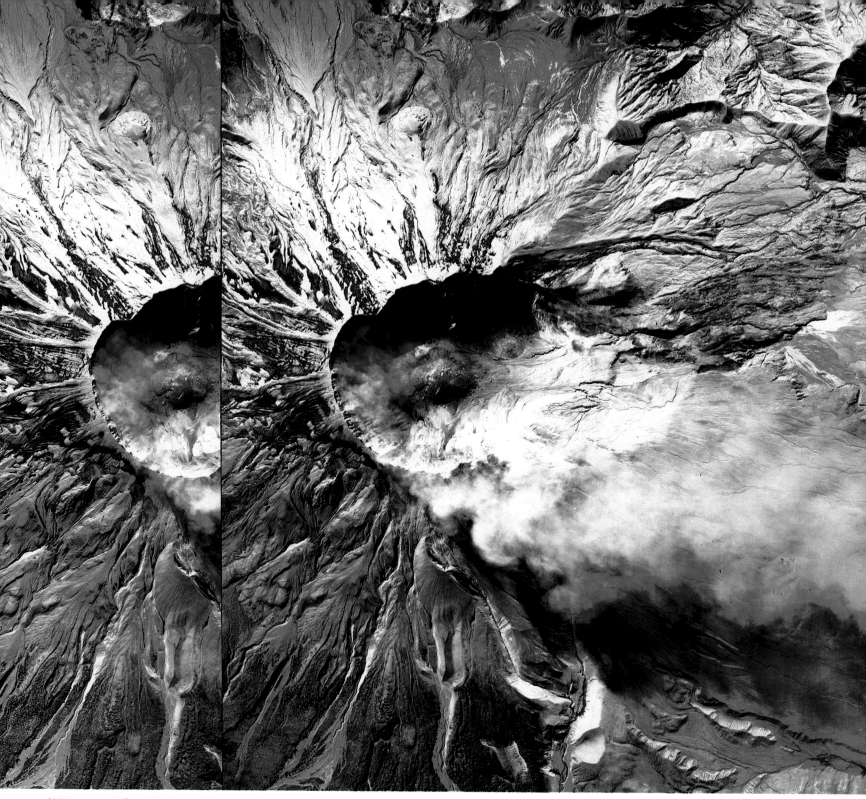

(U.S. Department of Agriculture)

Color infrared photo

0   0.5 miles

7. Mount St. Helens, Washington

Volcanic eruptions associated with converging plate margins are distinctive because the viscous, silica-rich magma produces violent eruptions of ash, ash flows, and thick andesitic lava flows. The result is a buildup of high, steep volcanoes, such as Mount St. Helens. The eruption on May 18, 1980, produced ash flows and mud flows (lahars), downed timber from the blast, and generated flooding.

a. Study the area in stereoscopic view and briefly describe the effects of the eruption.

b. Study the dome inside the crater and explain its origin.

c. How many individual lava flows can you identify? How do they differ from the plateau of flood basalts shown on page 214?

d. Draw an idealized cross section showing the internal structure of Mount St. Helens. Why is this type of volcano called a *stratovolcano,* or a *composite volcano?*

235

8. The Alpine–Himalayan Mountain Belt

The Alpine–Himalayan Mountain Belt extends from the Atlas Mountains in North Africa through southern Europe, the Middle East, and into Southeast Asia. It is the result of the collision of the African and Indian plates with the European and Asian plates. Throughout this great mountain belt, almost every kind of deformational structure that can be produced by continent–continent convergence is present. These tectonic plates continue to move, causing active crustal deformation, magmatic activity, and seismicity.

a. Draw the axis of the mountain belt on a piece of tracing paper. How does the trend of this line compare with that of the Cordilleran Mountain belt that extends from Alaska to the tip of South America?

b. Refer to the world map on pages 198–199. What is the relationship between the Himalayas and the islands of Sumatra and Java? What tectonic feature connects the two areas?

c. What is the origin of the lowlands (the Ganges and Indus river plain) in northern India and Pakistan?

d. What is the origin of the lowland Tigris and Euphrates rivers in Iraq?

e. How is the diverging motion of the Red Sea transformed into the converging motion of the folded mountains in Turkey?

f. There are no deep earthquakes in the Himalaya region. Explain why.

236

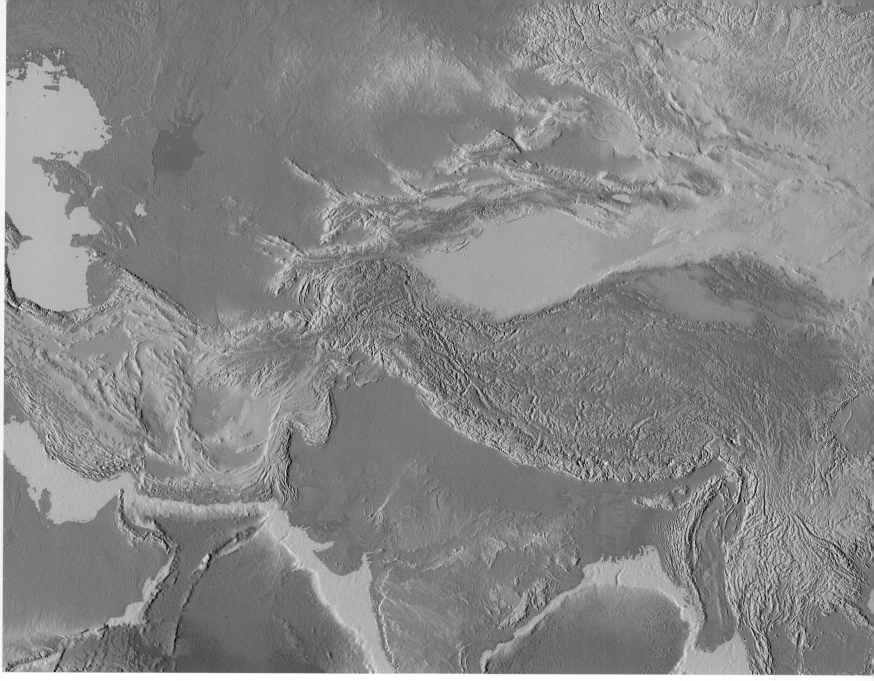

(Ken Perry, Chalk Butte, Inc.)

PLUMES AND HOTSPOTS

OBJECTIVE

To become acquainted with the nature and distribution of hotspots and the structural and topographic features they produce.

MAIN CONCEPT

Intense volcanic activity in the interior of tectonic plates is believed to be a surface manifestation of mantle plumes—long, narrow columns of hot material that slowly rise from the deep mantle.

SUPPORTING IDEAS

1. The surface expression of mantle plumes is intense basaltic volcanism (flood basalts) and broad, domal upwarps.
2. Plumes are generated deep in the mantle. Their location is fixed and independent of the overlying shifting plates.
3. Basaltic magma is generated because of decompression melting of the rising hot mantle plume.
4. A plume rising beneath oceanic lithosphere generates a large plateau of flood basalt. A narrow chain of volcanic islands may form as the tectonic plate moves over the plume.
5. Plumes rising beneath a continent produce a regional domal upwarp in the crust and eruption of floods of basalt.

CHARACTERISTICS OF HOTSPOTS

Although the most intense tectonic activity on Earth occurs along plate margins, local, isolated volcanic activity occurs where material from deep in the mantle upwells in long, narrow, vertical columns called mantle plumes. Where the plumes reach the surface, they form *hotspots*—areas that have abnormally high heat flow, volcanic activity, and broad crustal upwarps. The height of the relief is a consequence of thermal dilation and extrusion of a massive supply of basalt. Some geologists think that mantle plumes originate at depths of at least 700 km, and perhaps as deep as 2700 km, which is at the core–mantle boundary. An important characteristic is that the position of a plume is relatively stationary and thus the plume is independent of the movement of the overlying lithospheric plates. Mantle plumes may be expressed at the surface in several different ways: (1) as oceanic plateaus, (2) as chains of islands and

seamounts, (3) as continental basalt plateaus, and (4) as chains of mafic/silicic igneous centers.

If a plume rises beneath the ocean crust, distinctive landforms are produced. Decompression melting occurs as the enlarged plume head rises, and the hot magma is extruded rapidly as a spasm of volcanic activity, resulting in great floods of basalt erupted from fissures. This rapid extrusion builds a broad plateau that may rise thousands of meters above the surrounding seafloor. The result is a segment of oceanic crust 25- to 40-km thick—as much as five times the thickness of normal oceanic crust. In some areas, more than 35 million cubic kilometers of basalt may be extruded—enough to cover the entire United States with a layer 5 meters thick.

An important feature is that the floods of lava bury the older oceanic crust with its magnetic stripes that originally formed at the oceanic ridge and obscure the systematic pattern that exists throughout the rest of the ocean floor. Eventually, heat from the plume

head is exhausted through prolonged volcanic activity, and the production of magma declines. The subsequent volcanic history is dominated by continual extrusion of lava generated in the long tail of the plume. Uplift and basaltic volcanism occur during this part of plume evolution, but extrusion is less than that occurring during the initial phase involving the plume head. As the tectonic plate continually moves over the plume's tail, a chain of volcanic islands and seamounts is formed.

If a plume develops beneath a continent, it may cause regional uplift and eruption of continental flood basalt. Rhyolitic caldera systems may also develop if continental crust is partially melted by hot basaltic magma from the plume. Continental crust is not strongly deformed above a mantle plume, but the lithosphere bends to form broad swells and troughs. This bending may trigger shallow earthquakes. Sometimes, continental rifting and the development of an ocean basin may follow the development of a new plume.

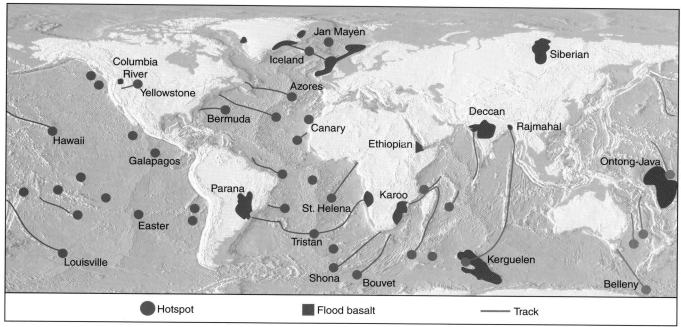

(Ken Perry, Chalk Butte, Inc.)

PROBLEMS

1. Global Distribution of Hotspots

With the help of the map above and the map at the end of the manual, locate the following features on the world map, pages 198–199.

a. Hotspots:

(1) Hawaii

(2) Easter Island

(3) Louisville

(4) Tristan

(5) St. Helens

How are these expressed at the surface?

b. Oceanic plateaus:

(1) Ontong Java

(2) Kerguelen

(3) Iceland

How are they expressed at the surface? What is the approximate area covered by each plateau?

c. Continental plateau basalts:

(1) Siberian

(2) Deccan

(3) Parana

(4) Karoo

(5) Ethiopian

(6) Columbia River

How are they expressed at the surface?

d. Study the location of hotspots, plateau basalts, and chains of islands and seamounts in the Atlantic Ocean. What evidence suggests that the initial phase of a rising plume produces great floods of basalt, and that chains of islands and seamounts result from the tail of plume activity?

e. Why are there so many remnants of flood basalts in or on the margins of the Atlantic and Indian oceans, and so few in the Pacific?

f. What is the ultimate fate of oceanic plateaus? Of seamount chains?

2. Ontong Java

a. Locate this area on the global map on pages 198–199. Briefly describe the physical characteristics (shape, size, height, etc.) of the Ontong Java plateau. Note that the ocean floor throughout most of the Pacific Ocean (shown in dark blue tones) is roughly 20,000 feet deep.

b. Calculate the volume of lava in the Ontong Java Plateau.

c. What does the buildup of such a great mass of basalt indicate concerning the stability of the position of the mantle plume that was the source of magma?

d. What does the volume of basalt indicate concerning the rate of extrusion?

e. How does the volume of basalt on Ontong Java compare with that of the great volcanoes on Mars (page 251)?

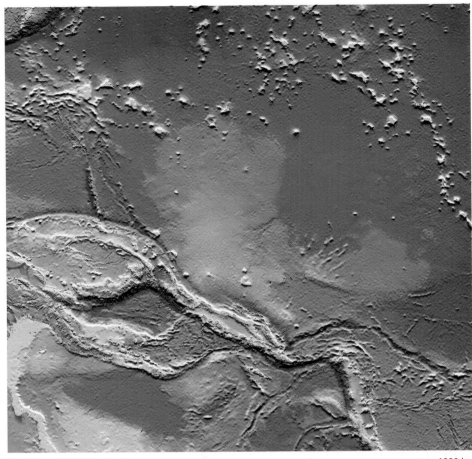

(Ken Perry, Chalk Butte, Inc.)

0   1000 km

3. Kerguelen

The image of the Kerguelen Plateau is distorted because of the map projection. It is not as large as it appears on this map. This is the same phenomenon seen in certain map projections, such as Mercator, that makes Greenland look larger than South America.

a. Locate this area on the map on pages 198–199. Compare the physical characteristics of the Kerguelen Plateau with the Ontong Java Plateau.

b. Note the margins of microcontinents, such as Madagascar and the Seychelles Islands, on the map on pages 198–199. They are composed of continental granitic crust rifted by faulting. How do the margins of Kerguelen, Ontong Java, and Iceland differ from fragments of continental crust?

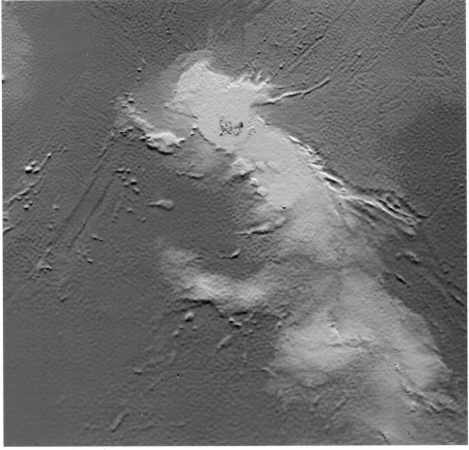

(Ken Perry, Chalk Butte, Inc.)

0   500 km

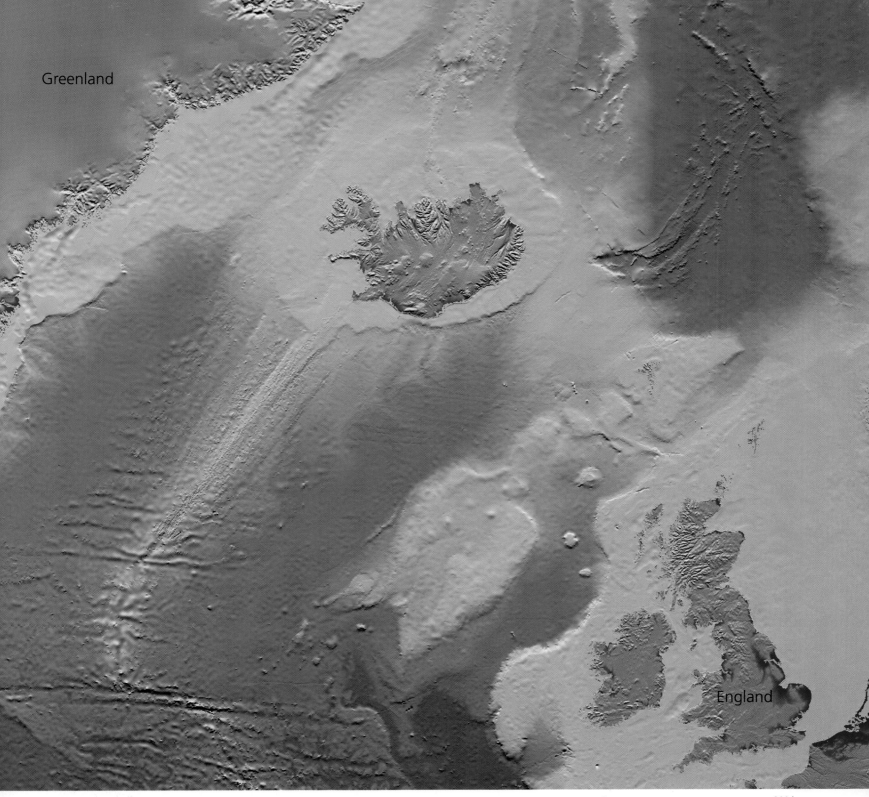

Greenland

England

(Ken Perry, Chalk Butte, Inc.)

0   200 km

4. Iceland

a. Draw arrows on the map to show plate motion.

b. Why is the oceanic ridge north and south of Iceland so high and rugged compared to that of the rest of the Atlantic?

c. What evidence supports the idea that a large plume is stationary beneath Iceland and that volcanic deposits from it have moved with the shifting plates?

d. Explain the origin of the long, narrow, high ridge that extends along the Mid-Atlantic Ridge south of Iceland.

e. Explain the origin of the oceanic plateau basalts northwest of Great Britain and near the southern margins of Greenland (expressed as very light tones of blue, because the top of the plateau rises almost to sea level).

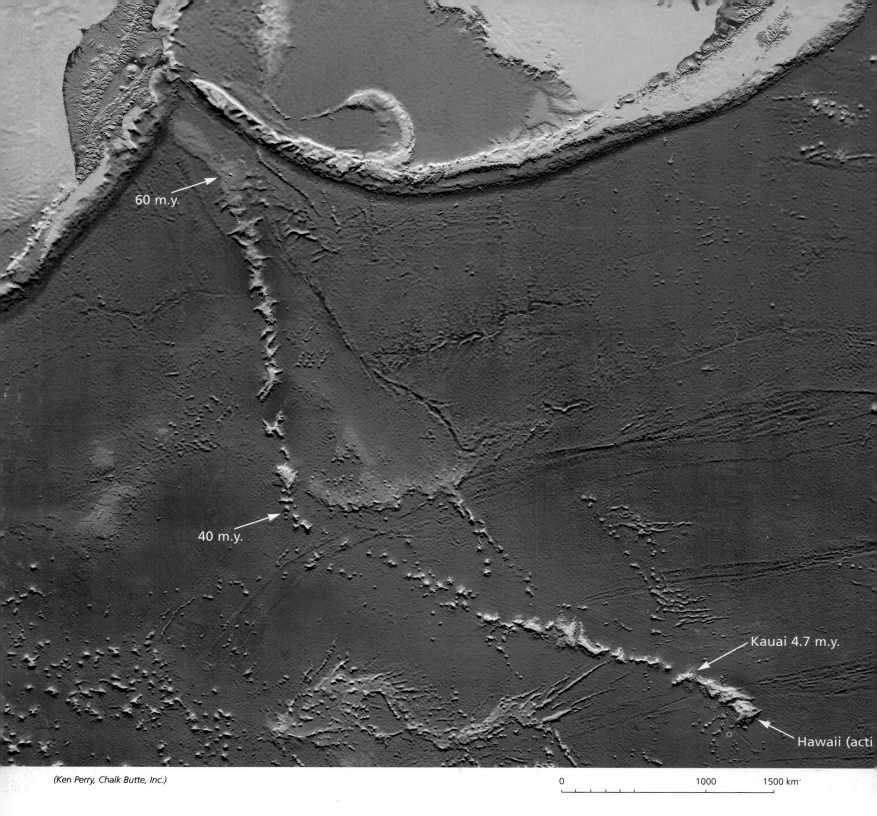

60 m.y.

40 m.y.

Kauai 4.7 m.y.

Hawaii (acti

(Ken Perry, Chalk Butte, Inc.)

0   1000   1500 km

5. Hawaii–Emperor Seamount Chain

This map shows the Hawaiian Islands and the Emperor Seamount chain and the age of several islands in millions of years. Kilauea and Mauna Loa, on the big island of Hawaii, are presently active. Loihi, a seamount south of Hawaii, is also active and will build a new island in the future.

a. What has been the rate of plate movement during the last 4.7 million years?

b. What has been the direction and rate of plate movement during the last 40 million years?

c. What has been the direction and rate of plate movement during the period from 40 million years to 60 million years ago?

d. Explain why the islands in the Hawaiian chain become progressively submerged as they recede from the hotspot, whereas those of the Emperor chain are all at about the same depth.

e. Could there once have been an oceanic plateau built by the Hawaiian hotspot?

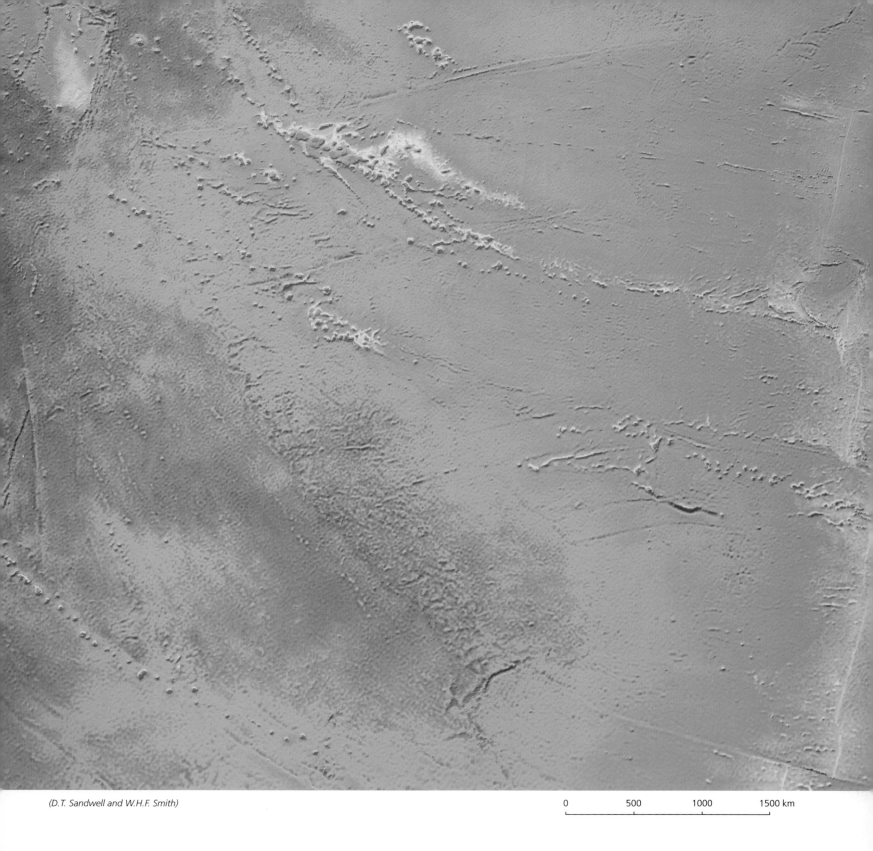

(D.T. Sandwell and W.H.F. Smith)

| 0 | 500 | 1000 | 1500 km |

6. Islands and Seamounts in the South Pacific

a. Locate this area on the map on pages 198–199. On the basis of islands and seamounts shown on this map, plot the position of the major hotspots in the South Pacific region.

b. Is there any indication from the chains of islands and seamounts

that the direction of plate motion has changed in the last 60 million years?

c. Plot the direction of plate movement based on "hotspot tracks."

7. Siberian Plateau Basalts

The basalts of Siberia form a large plateau covering nearly 400 million square miles. They were extruded during Permian–Triassic time and represent one of the largest outpourings of lava in geologic history.

> **a.** Study this image and the regional map on pages 198–199. Is the Siberian Plateau related to any rift system or was it formed from an isolated plume?
>
> **b.** Based on the drainage patterns superposed on the basalts, where would be the highest area of the plateau and, therefore, the likely center of the hotspot? Mark this on the map.
>
> **c.** How have the basalts influenced the regional drainage systems?
>
> **d.** How does the size of the Siberian Plateau compare with the size of oceanic plateaus?

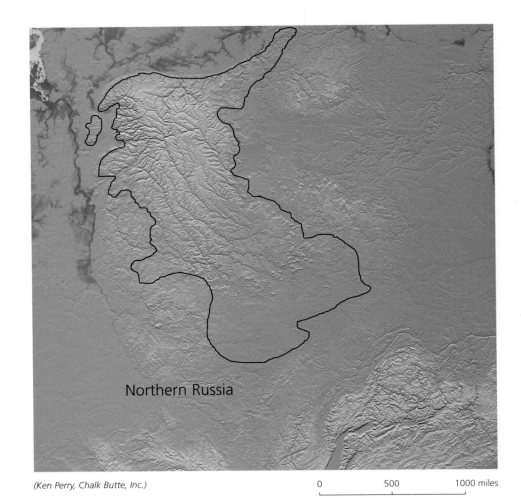

Northern Russia

(Ken Perry, Chalk Butte, Inc.)

0   500   1000 miles

8. Parana Plateau, South America

> **a.** Study the plateau on this map and on pages 198–199. The basalts are approximately 200 million years old. Were the basalts produced from an isolated plume or from a plume associated with continental rifting? Explain the basis for your answer.
>
> **b.** What influence did the formation of the Parana basalts have on regional drainage of South America?
>
> **c.** What evidence on this image indicates that the basalts are relatively old?

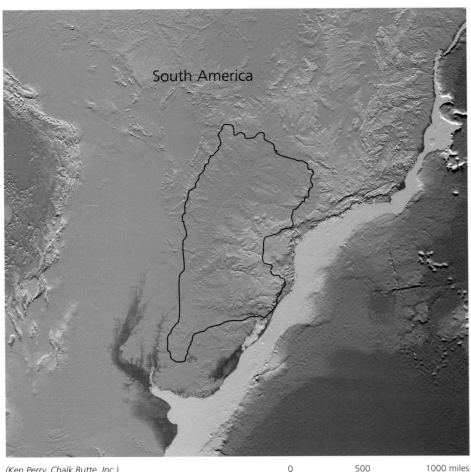

South America

(Ken Perry, Chalk Butte, Inc.)

0   500   1000 miles

9. Deccan Plateau, India

a. Is the tectonic setting of the hotspot that produced the basalts of the Deccan Plateau similar to the hotspot that produced the Siberian Plateau or the Parana Plateau? Explain the basis for your answer.

b. In which of the following areas would you expect to find a rock sequence similar to that of the Deccan Plateau: (1) Siberian, (2) Karoo, (3) Ethiopian, (4) Iceland? Explain the basis for your answer.

c. During the time of its formation, would you expect the Deccan Plateau to be similar to (1) Siberian, (2) Ontong Java, (3) Ethiopian, or (4) Parana? Explain the basis for your answer.

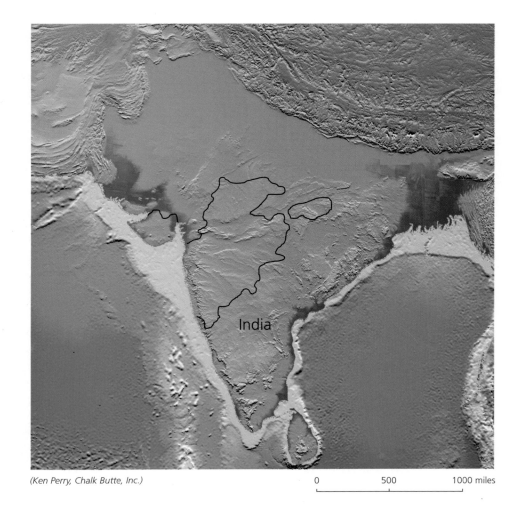

(Ken Perry, Chalk Butte, Inc.)

0   500   1000 miles

10. Ethiopian Plateau

The Ethiopian Plateau is one of the highest and most rugged parts of Africa.

a. What relationship does the Ethiopian Plateau have with the major tectonic plates?

b. Judging from the elevation of the top of the plateau, is the sequence of basalts thicker or thinner than that of the other continental plateau basalts?

c. What processes have combined to dissect the plateau with deep valleys?

d. If present tectonic processes continue, what will happen to the Ethiopian Plateau?

e. Draw a schematic cross section across the Ethiopian Plateau showing the topography and distribution of basalt and the faults you think are present.

f. How are the plateau basalts related to a triple junction?

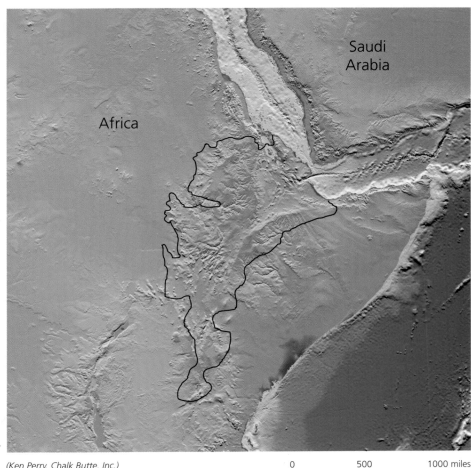

(Ken Perry, Chalk Butte, Inc.)

0   500   1000 miles

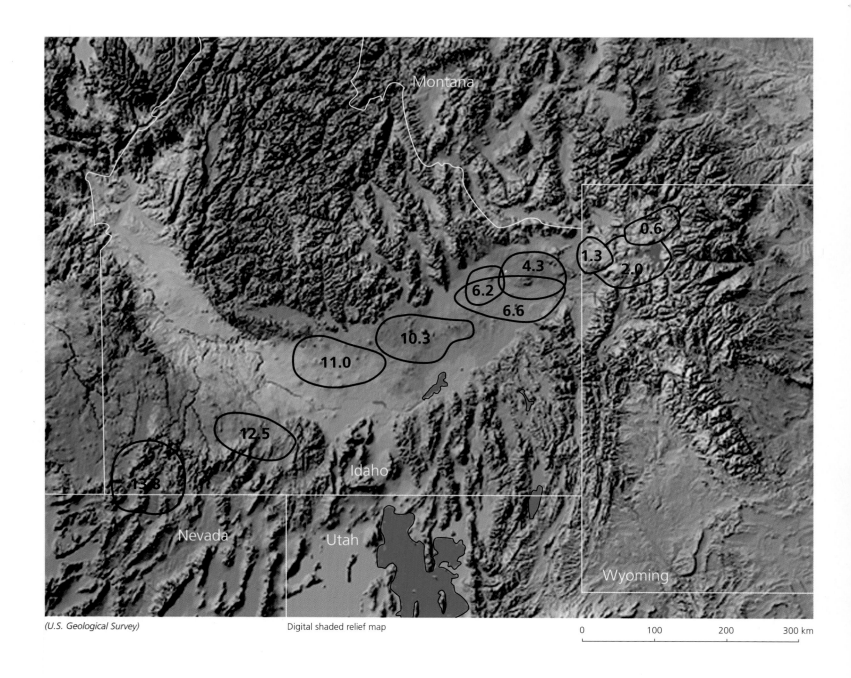

(U.S. Geological Survey)   Digital shaded relief map

0   100   200   300 km

11. Snake River Plain

The Snake River Plain is part of the
Columbia River flood basalt system.
Compared to other basalt plateaus, it is
very young and is still active. This map
shows the distribution of calderas in the
Snake River Plain, with the numbers
showing the age of rock formed in each
area in millions of years.

a. What is the direction and rate of
plate movement over the hotspot?

b. What area will be over the
plume 10 million years from now?

EXERCISE 23   PLANETARY GEOLOGY

■ OBJECTIVE

To recognize the major types of landforms on the various planetary bodies in the solar system and to understand the processes that formed them.

■ MAIN CONCEPT

Space photographs of the planetary bodies in the solar system show that a planet's rocks and surface features contain a record of the events that result from the specific energy systems operating on or within it. With this new information, we can compare the geology of Earth with the geology of other planetary bodies, and identify those processes that are fundamental to all planets and those that are unique or of special importance to a specific body.

■ SUPPORTING IDEAS

1. The sequence of events in a planet's geologic history can be determined by using the fundamental principles of geology, such as superposition and cross-cutting relationships.
2. The impact of a meteorite produces a new landform, known as a *crater,* and a new rock body consisting of the blanket of *ejecta* that surrounds the crater.
3. The surfaces of many planetary bodies have been modified by processes resulting from internal heat.
4. The most common expressions of internal heat in a planetary body are volcanic extrusions, folding, and faulting.

The dominant surface features on most planetary bodies are impact craters, which were probably produced during the last phases of planetary accretion.

In simple terms, the impact of a meteorite produces two geologic features: (1) an impact crater and (2) a body of fragmented rock (an ejecta blanket) thrown out of the crater by the impact.

The large-scale deformation that produces the structure of the lithosphere of a planetary body is the result of tectonism. Internal heat is the driving force, but the response (flowing, buckling, folding, or fracturing) depends on the strength of the lithosphere.

On the basis of the superposition of ejecta or volcanic material, or both, the relative sequence of many events in the geologic history of a planetary body can be determined. An example of the landforms on the Moon is shown in Figure 23.1.

PROBLEMS

To solve the following problems, remember that the age of a planetary surface is reflected by its relative number of impact structures. Old surfaces have been subjected to long periods of bombardment and are saturated with craters. Younger surfaces have fewer craters.

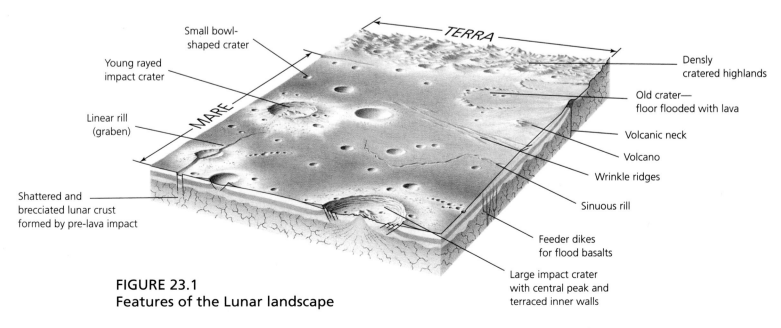

FIGURE 23.1
Features of the Lunar landscape

247

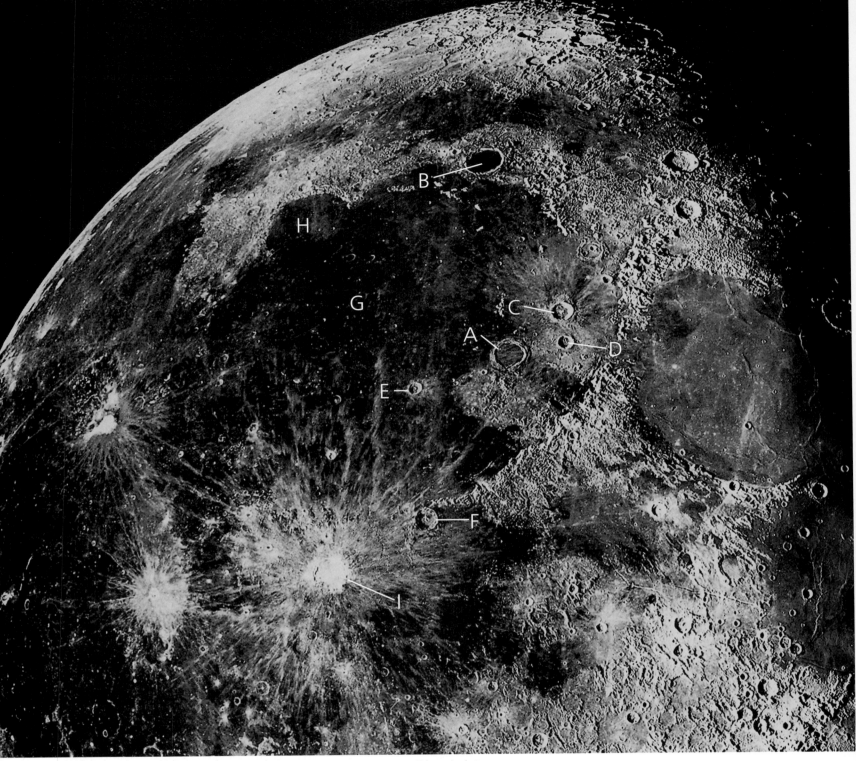

(Courtesy of the Jet Propulsion Lab, California Institute of Technology)   Telescopic photo

Parts of the lunar surface have been photographed stereoscopically and can be studied in the same way that we study stereoscopic photographs of Earth.

1. Mare Imbrium—Moon

a. What evidence indicates that the craters Archimedes (A) and Plato (B), together with their ejecta blankets, are older than Aristillus (C), Autolycus (D), Timocharis (E), and Eratosthenes (F)?

b. Note that the Imbrium Basin (G) is considered to be a multiring crater. What evidence supports this conclusion?

c. What evidence can you find that the crater Sinus Iridum (H) and its ejecta blanket are younger than the ejecta of the multiring crater that forms the Imbrium Basin?

d. What evidence suggests that Copernicus (I) and its ejecta blan-

ket are the youngest features in the area?

e. Arrange the following features in proper chronologic sequence:

- Eratosthenes (ejecta blanket)
- Mare Imbrium (basalt flows)
- Archimedes (ejecta)
- Copernicus (ejecta)

2. Mare Imbrium—Moon

a. Study the ejecta blanket of the large crater in the lower part of the photograph. Is this body of rock older or younger than the basalts of the surrounding maria? Explain.

b. A number of individual lava flows can be seen in this photograph. Trace their margins and try to determine their relative age. On the basis of your mapping, where do you think the center of extrusion was?

c. What types of volcanic eruptions have occurred in this area?

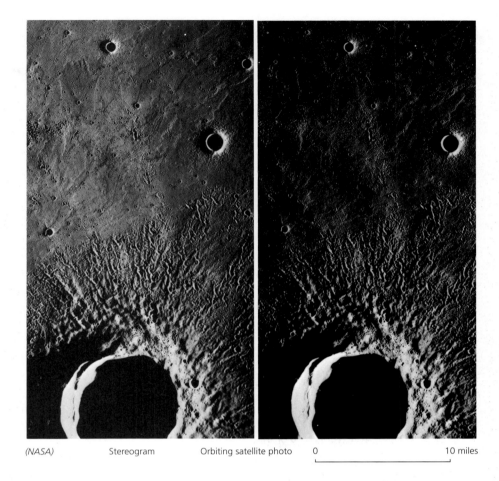

(NASA)   Stereogram   Orbiting satellite photo   0 _____ 10 miles

3. Mare Imbrium—Moon

a. Name the features identified on the photograph by the numbers 1, 2, 3, and 4.

b. Are the craters older or younger than the smooth lava flows?

c. Map the extent of the ejecta blanket associated with the largest crater.

d. What was the most recent geologic event in this area?

e. Where are the oldest rocks exposed?

f. How do the younger craters in this area, and in the photograph on page 248, differ from the older craters?

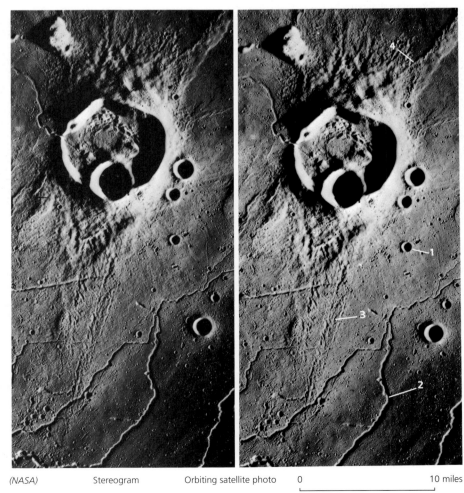

(NASA)   Stereogram   Orbiting satellite photo   0 _____ 10 miles

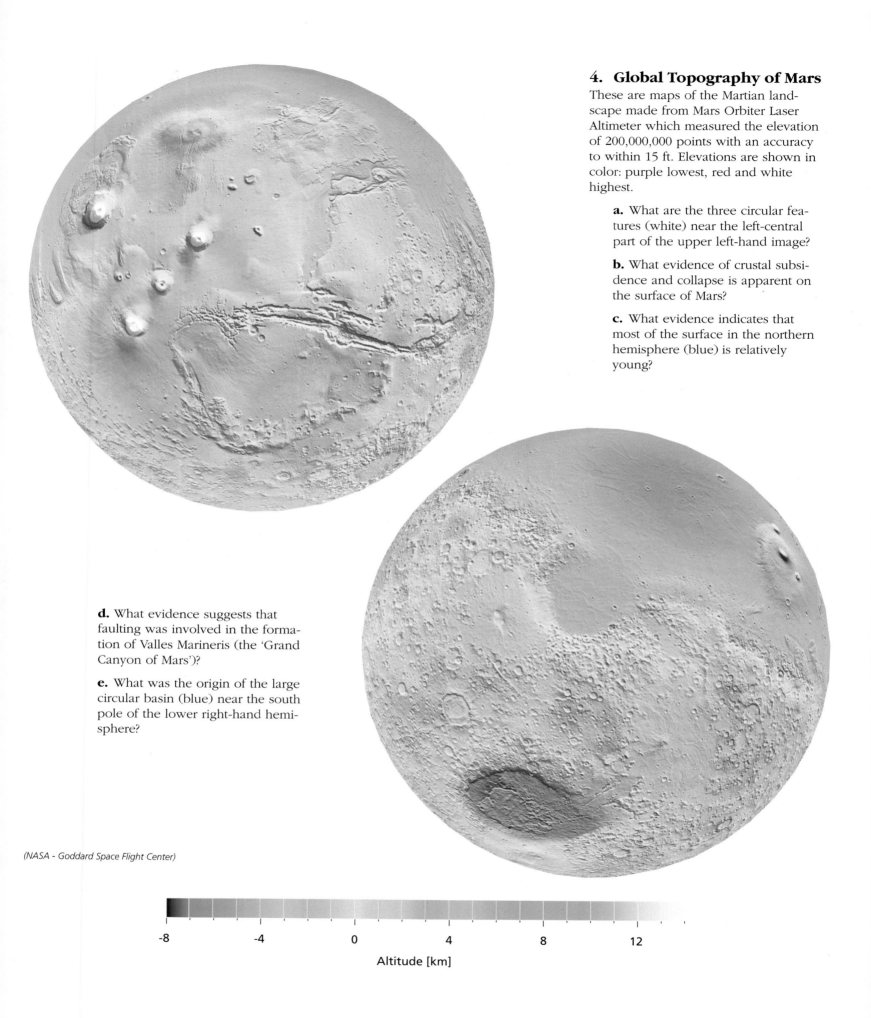

4. Global Topography of Mars

These are maps of the Martian land-scape made from Mars Orbiter Laser Altimeter which measured the elevation of 200,000,000 points with an accuracy to within 15 ft. Elevations are shown in color: purple lowest, red and white highest.

a. What are the three circular features (white) near the left-central part of the upper left-hand image?

b. What evidence of crustal subsidence and collapse is apparent on the surface of Mars?

c. What evidence indicates that most of the surface in the northern hemisphere (blue) is relatively young?

d. What evidence suggests that faulting was involved in the formation of Valles Marineris (the 'Grand Canyon of Mars')?

e. What was the origin of the large circular basin (blue) near the south pole of the lower right-hand hemisphere?

(NASA - Goddard Space Flight Center)

-8   -4   0   4   8   12

Altitude [km]

5. Close-up of Mars

The Mars Global Surveyor provided unprecedented close-up views of the planet with a camera and suite of remote-sensing instruments. By the year 2000 some 80,000 images, 50 times as detailed as any previously taken, revealed objects on Mars' surface five feet across, and indicated that Mars is presently much more active than previously thought.

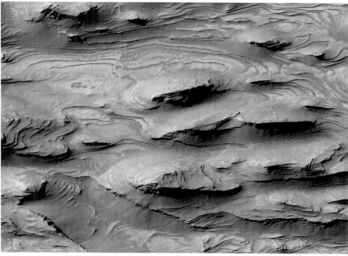

0   500 Feet

B. Rock Layers in Valles Marineris

A sequence of rock layers nearly 3000 ft thick is exposed within the walls of Valles Marineris. There are over 100 beds in this area, each with the same thickness (30 ft).

1. List three possible ways in which the sequence of strata could have formed.

2. Layers indicate change. What does the uniform pattern seen here (beds of similar thickness and physical properties) suggest about the geologic progress which formed the sequence of layers?

3. The terrain has been eroded into elongate elliptical ridges and troughs. What agent most likely produced these features?

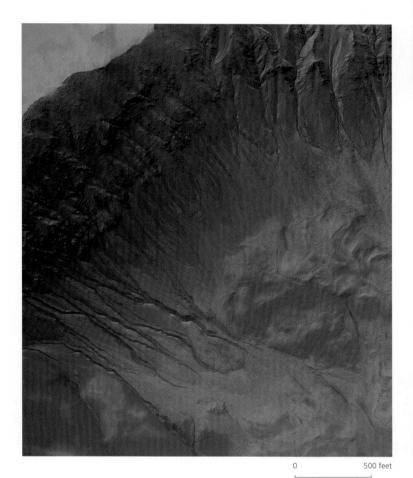

0   500 feet

A. Stream Erosion

This is a view of part of a crater rim comprising an area of 2.2 X 3.3 miles.

1. What evidence indicates that running water has recently flowed on Mars?

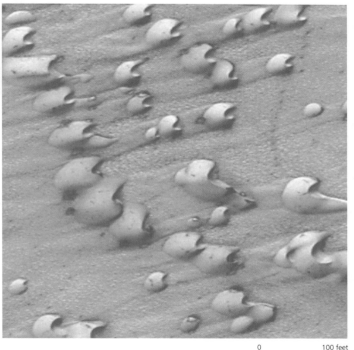

0   100 feet

(U.S. Geological Survey)   Shaded relief map

C. Sand Dunes

1. What type of dunes occur in this area?

2. What do these dunes tell us about sand supply, wind velocity, and wind direction?

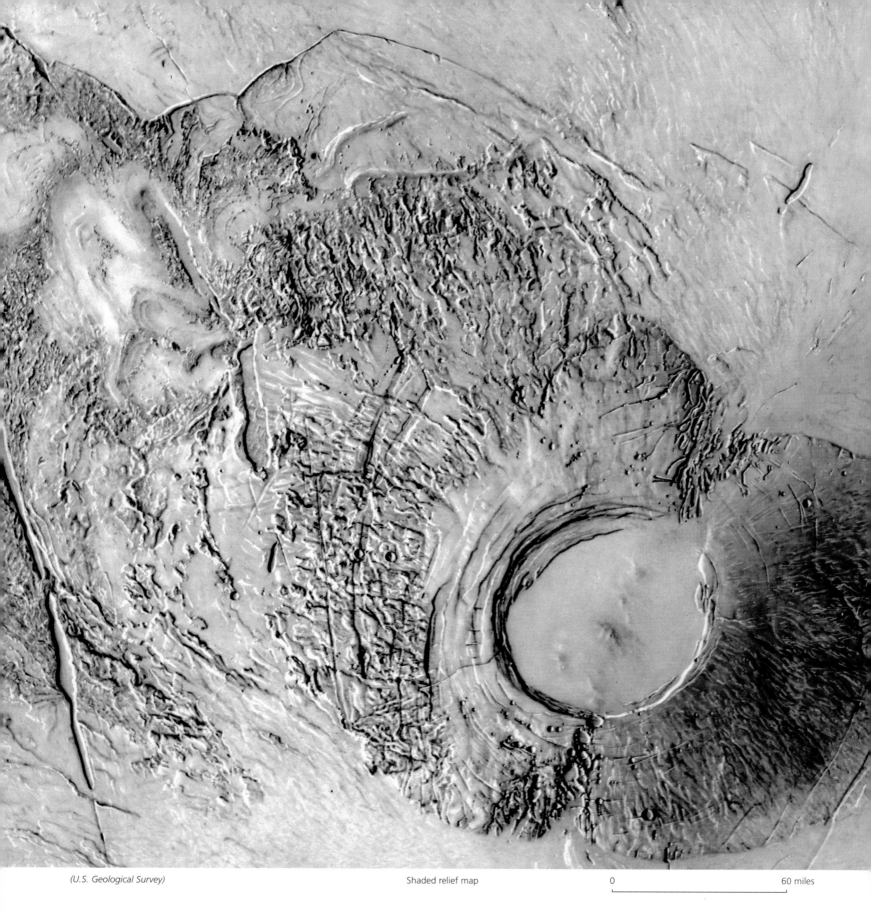

Shaded relief map

0   60 miles

6. Phoenicis Lacus Region—Mars

a. What features indicate that the large circular structure is a volcano and not an impact crater?

b. How has the volcano been modified by tectonic processes?

c. What evidence indicates that the entire surface shown on this map is relatively young?

d. What evidence indicates that a period of fissure eruptions of flood basalts occurred before the formation of the large volcano?

e. What are the youngest structural features in this area? What types of events do they record?

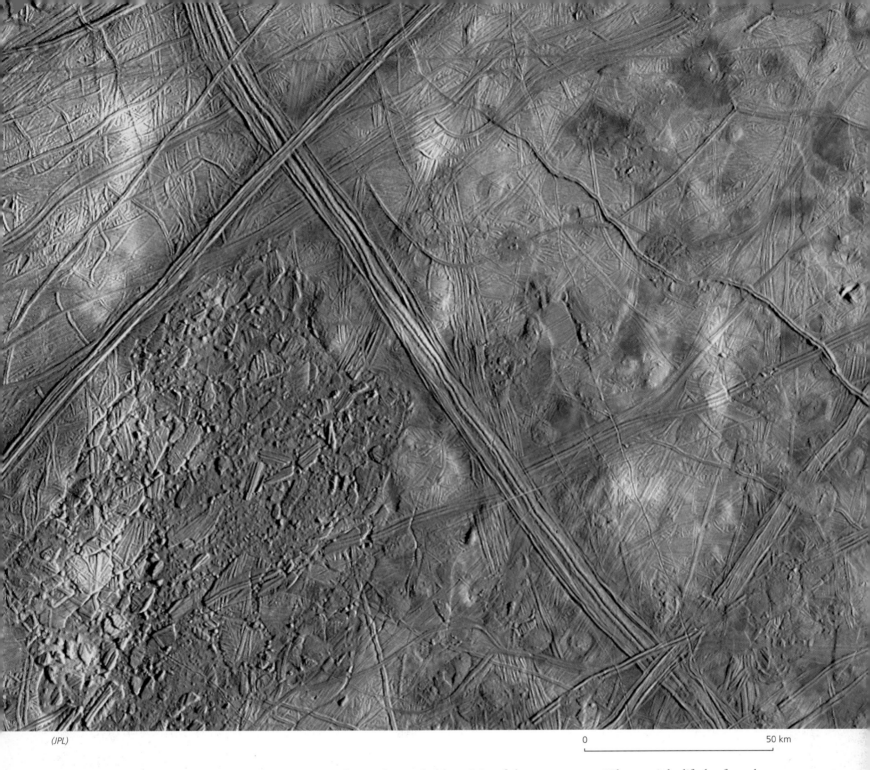

(JPL)

0   50 km

7. Europa

The surface texture, color, and shading indicate that various regions on Europa are made of different materials, formed in unique ways at different times. Icy plains are shown in blue tones: coarse-grained ice is dark blue; fine-grained ice is light blue. Reddish-brown areas are surface materials other than pure ice resulting from geologic activity along the fracture zones. White areas are parts of ray material resulting from an impact crater far beyond this scene.

a. Study the long linear structures (fractures), which are emphasized by hues of brown and orange.

What is the probable origin of the ridges and valleys parallel to the axis of the structure?

b. What features of Earth are similar to these structures?

c. What is the oldest terrain in this area of Europa? How did you determine this?

d. Where is the youngest terrain in this area?

e. How many periods of deformation can you recognize?

f. What other features on Earth are similar to those in this part of Europa?

g. Where might life be found on Europa?

h. What evidence indicates that the surface of Europa is younger than that of our moon?

i. Study the "jumbled" terrain, a disorganized area consisting of angular blocks and fragments. What type of motion would produce such terrain? How would you describe the physical properties of the material beneath the surface (solid, slushy, liquid, etc.)?

253

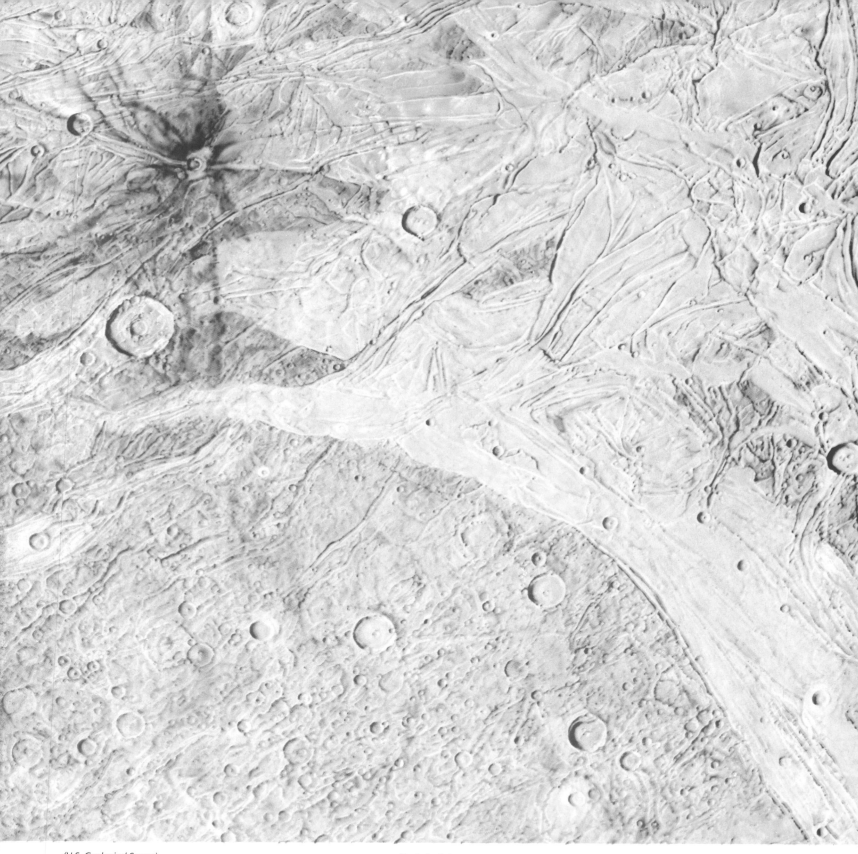

(U.S. Geological Survey)   Shaded relief map   0 _____ 60 miles

8. Surface Features of Ganymede

a. Based on the relative number of craters, which terrain type would you judge to be older: (1) the dark, fragmented terrain or (2) the light, grooved terrain?

b. What are the youngest features on Ganymede?

c. What evidence indicates that the crust of Ganymede has moved and shifted about?

d. Explain the nature and origin of the "volcanism" on Ganymede.

e. Construct a preliminary geologic column showing the relative age of the major geologic events in this area.

254

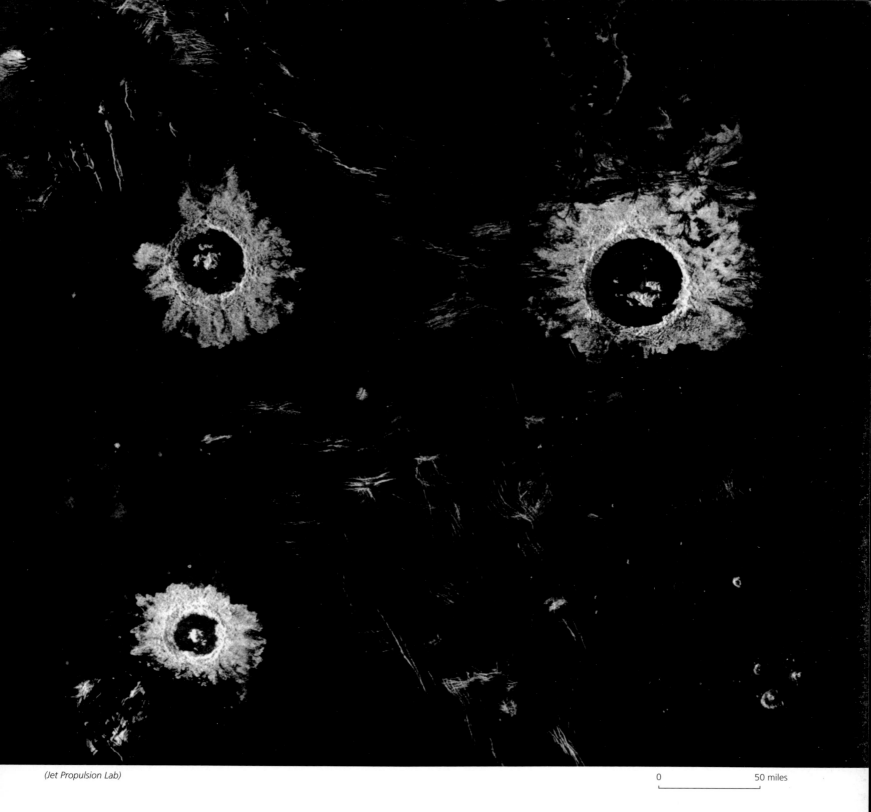

(Jet Propulsion Lab)

0   50 miles

9. Surface Features of Venus

This radar image shows an area approximately 342 mi wide and 311 mi long. The major landforms include impact craters, faults, plains, small circular domes, and lava flows. The surface of Venus is 890 °F, hot enough to melt lead.

a. Compare and contrast the craters on Venus with those on the Moon (page 249). Why does the material ejected from the impact crater on Venus occur in distinct lobes?

b. How many major fracture systems occur in this area? What is the type of faulting (normal, thrust, or strike-slip)? What surface fractures support your answer?

c. What is the probable origin of the small domes shown in the southeast corner of the image? (The domes range in diameter from 0.6 to 7 mi, and some have summit depressions.)

d. Note the lava flows, characterized by a faint, light border, that appear to emanate from the impact craters. What might be the origin of these flows?

e. What is the probable origin of the smooth plain on which all of the other features occur?

f. Judging from the number of craters on this surface, would you consider the plains to be a relatively young, intermediate, or old surface?

10. Circular Domes—Venus

a. Explain the origin of the domes.

b. Why is a distinctive fracture system developed on each dome? What caused the fractures?

c. Are the domes older or younger than the fractures that cut the surrounding plains?

d. How do the domes differ from similar features formed on Earth?

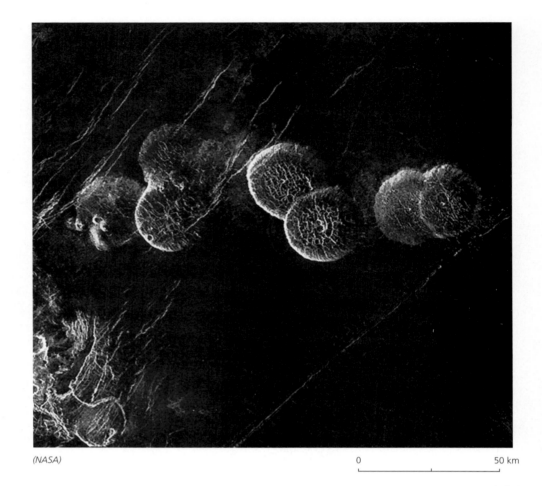

(NASA)

0   50 km

11. Spiderweb Fractures— Venus

a. This fracture pattern, so far, has been found only on Venus. The centers of the circular structures are lower than the surrounding areas. Study the fracture patterns and develop several working hypotheses that explain their origin.

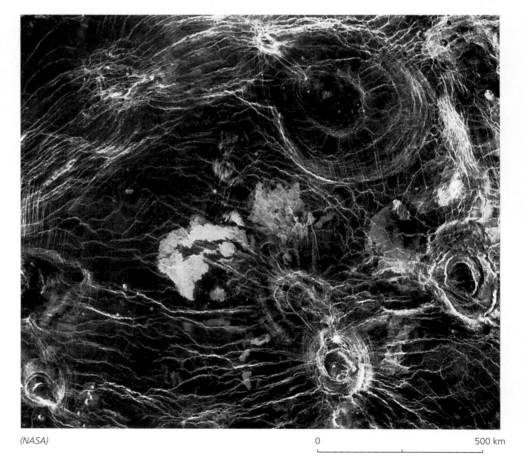

(NASA)

0   500 km

TOPOGRAPHIC MAP SYMBOLS

BOUNDARIES

National . . . .

State or territorial . . . .

County or equivalent . . . .

Civil township or equivalent . . . .

Incorporated-city or equivalent . . . .

Park, reservation, or monument . . . .

Small park . . . .

LAND SURVEY SYSTEMS

U.S. Public Land Survey System:

Township or range line . . . .

Location doubtful . . . .

Section line . . . .

Location doubtful . . . .

Found section corner; found closing corner . . . . . .

Witness corner; meander corner . . . .

Other land surveys:

Township or range line . . . .

Section line . . . .

Land grant or mining claim; monument . . . .

Fence line . . . .

ROADS AND RELATED FEATURES

Primary highway . . . .

Secondary highway . . . .

Light duty road . . . .

Unimproved road . . . .

Trail . . . .

Dual highway . . . .

Dual highway with median strip . . . .

Road under construction . . . .

Underpass; overpass . . . .

Bridge . . . .

Drawbridge . . . .

Tunnel . . . .

BUILDINGS AND RELATED FEATURES

Dwelling or place of employment: small; large . . .

School; church . . . .

Barn, warehouse, etc.: small; large . . . .

House omission tint . . . .

Racetrack . . . .

Airport . . . .

Landing strip . . . .

Well (other than water); windmill . . . .

Water tank: small; large . . . .

Other tank: small; large . . . .

Covered reservoir . . . .

Gaging station . . . .

Landmark object . . . .

Campground; picnic area . . . .

Cemetery: small; large . . . . Cem

RAILROADS AND RELATED FEATURES

Standard gauge single track; station . . . .

Standard gauge multiple track . . . .

Abandoned . . . .

Under construction . . . .

Narrow gauge single track . . . .

Narrow gauge multiple track . . . .

Railroad in street . . . .

Juxtaposition . . . .

Roundhouse and turntable . . . .

TRANSMISSION LINES AND PIPELINES

Power transmission line: pole; tower . . . .

Telephone or telegraph line . . . .

Aboveground oil or gas pipeline . . . .

Underground oil or gas pipeline . . . .

CONTOURS

Topographic:

Intermediate . . . .

Index . . . .

Supplementary . . . .

Depression . . . .

Cut; fill . . . .

Bathymetric:

Intermediate . . . .

Index . . . .

Primary . . . .

Index Primary . . . .

Supplementary . . . .

MINES AND CAVES

Quarry or open pit mine . . . .

Gravel, sand, clay, or borrow pit . . . .

Mine tunnel or cave entrance . . . .

Prospect; mine shaft . . . .

Mine dump . . . .

Tailings . . . .

SURFACE FEATURES

Levee . . . .

Sand or mud area, dunes, or shifting sand . . . .

Intricate surface area . . . .

Gravel beach or glacial moraine . . . .

Tailings pond . . . .

VEGETATION

Woods . . . .

Scrub . . . .

Orchard . . . .

Vineyard . . . .

Mangrove . . . .

MARINE SHORELINE

Topographic maps:

Approximate mean high water . . . .

Indefinite or unsurveyed . . . .

Topographic-bathymetric maps:

Mean high water . . . .

Apparent (edge of vegetation) . . . .

COASTAL FEATURES

Foreshore flat . . . .

Rock or coral reef . . . .

Rock bare or awash . . . .

Group of rocks bare or awash . . . .

Exposed wreck . . . .

Depth curve; sounding . . . .

Breakwater, pier, jetty, or wharf . . . .

Seawall . . . .

BATHYMETRIC FEATURES

Area exposed at mean low tide; sounding datum . .

Channel . . . .

Offshore oil or gas: well; platform . . . .

Sunken rock . . . .

RIVERS, LAKES, AND CANALS

Intermittent stream . . . .

Intermittent river . . . .

Disappearing stream . . . .

Perennial stream . . . .

Perennial river . . . .

Small falls; small rapids . . . .

Large falls; large rapids . . . .

Masonry dam . . . .

Intermittent lake or pond . . . .

Dry lake . . . .

Narrow wash . . . .

Wide wash . . . .

Canal, flume, or aqueduct with lock . . . .

Elevated aqueduct, flume, or conduit . . . .

Aqueduct tunnel . . . .

Water well; spring or seep . . . .

GLACIERS AND PERMANENT SNOWFIELDS

Contours and limits . . . .

Form lines . . . .

SUBMERGED AREAS AND BOGS

Marsh or swamp . . . .

Submerged marsh or swamp . . . .

Wooded marsh or swamp . . . .

Submerged wooded marsh or swamp . . . .

Rice field . . . .

Land subject to inundation . . . .

Greenland
Iceland
Brooks Range
Bering Sea
Aleutian Islands
Cordilleran Mountains
North America
Juan de Fuca Ridge
Cascade Mts.
Emperor Seamounts
Appalachian Mts.
Mid Atlantic Ridge
Hawaii
Caribbean Sea
Amazon River
South America
Andes Mountains
Tuamotu Isls.
Tonga Trench
East Pacific Rise
Parana Plat.
Macdonald Isl.
Easter Isl.
Alpin fault
New Zealand
Scotia Arc